Jan Bauer

Origins of Non-Ideal Current–Voltage Characteristics of Si Solar Cells

Jan Bauer

Origins of Non-Ideal Current–Voltage Characteristics of Si Solar Cells

Analyses of the impact of shunts, breakdown sites, and defects on multicrystalline silicon solar cells

Südwestdeutscher Verlag für Hochschulschriften

Impressum / Imprint
Bibliografische Information der Deutschen Nationalbibliothek: Die Deutsche Nationalbibliothek verzeichnet diese Publikation in der Deutschen Nationalbibliografie; detaillierte bibliografische Daten sind im Internet über http://dnb.d-nb.de abrufbar.
Alle in diesem Buch genannten Marken und Produktnamen unterliegen warenzeichen-, marken- oder patentrechtlichem Schutz bzw. sind Warenzeichen oder eingetragene Warenzeichen der jeweiligen Inhaber. Die Wiedergabe von Marken, Produktnamen, Gebrauchsnamen, Handelsnamen, Warenbezeichnungen u.s.w. in diesem Werk berechtigt auch ohne besondere Kennzeichnung nicht zu der Annahme, dass solche Namen im Sinne der Warenzeichen- und Markenschutzgesetzgebung als frei zu betrachten wären und daher von jedermann benutzt werden dürften.

Bibliographic information published by the Deutsche Nationalbibliothek: The Deutsche Nationalbibliothek lists this publication in the Deutsche Nationalbibliografie; detailed bibliographic data are available in the Internet at http://dnb.d-nb.de.
Any brand names and product names mentioned in this book are subject to trademark, brand or patent protection and are trademarks or registered trademarks of their respective holders. The use of brand names, product names, common names, trade names, product descriptions etc. even without a particular marking in this works is in no way to be construed to mean that such names may be regarded as unrestricted in respect of trademark and brand protection legislation and could thus be used by anyone.

Coverbild / Cover image: www.ingimage.com

Verlag / Publisher:
Südwestdeutscher Verlag für Hochschulschriften
ist ein Imprint der / is a trademark of
AV Akademikerverlag GmbH & Co. KG
Heinrich-Böcking-Str. 6-8, 66121 Saarbrücken, Deutschland / Germany
Email: info@svh-verlag.de

Herstellung: siehe letzte Seite /
Printed at: see last page
ISBN: 978-3-8381-2865-8

Zugl. / Approved by: Halle, MLU, Diss., 2009

Copyright © 2012 AV Akademikerverlag GmbH & Co. KG
Alle Rechte vorbehalten. / All rights reserved. Saarbrücken 2012

„Atomkraft geht nicht. Im 21. Jahrhundert kommt der Strom aus Solarzellen."
"Nuclear power doesn't work. In the 21st century power comes from solar cells."

Ludwig Bölkow (1912 - 2003), Ingenieur (engineer)

Contents

Preface   v

1 Solar Cells   1
  1.1 Production of Multicrystalline Silicon Solar Cells 1
  1.2 Established Theory of Solar Cell I–V Characteristics 4
    1.2.1 One-Diode Model 4
    1.2.2 Two-Diode Model 5
  1.3 Breakdown Mechanisms in p-n Junctions 8
    1.3.1 Avalanche Breakdown 8
    1.3.2 Zener Effect 8
  1.4 Real I–V Characteristics 9
    1.4.1 Forward I–V Characteristics 9
    1.4.2 Previous Models for the Recombination Current 11
    1.4.3 Reverse I–V Characteristics 12

2 Measurement Methods   15
  2.1 Lock-in Thermography 15
    2.1.1 Imaging of Physical Parameters of Forward Biased Solar Cells 18
    2.1.2 Imaging of Physical Parameters of Reverse-Biased Solar Cells 21
  2.2 Luminesence Methods 26
  2.3 Electron Microscopy Methods 28
    2.3.1 SEM Techniques 28
    2.3.2 TEM 28
  2.4 Infrared Microscopy & Electrical Measurements 29

3 I–V Characteristics of Solar Cells   31
  3.1 Results: Forward I–V Characteristics 31
    3.1.1 Linear Shunts 32
    3.1.2 SiC Filaments 34
    3.1.3 Growth of SiC Filaments in Block-Cast mc-Si 38
    3.1.4 Non-Linear Shunts 46
  3.2 Results: Reverse I–V Characteristics 50
    3.2.1 The Term "Pre-Breakdown" 50
    3.2.2 Early Pre-Breakdown in Solar Cells: Linear Region 53
    3.2.3 Pre-Breakdown in Solar Cells: Defect-Induced Breakdown 54
    3.2.4 Hard Pre-Breakdown: Avalanche Breakdown in Solar Cells 62
    3.2.5 Comparision of Pre-Breakdown I–V Characteristics 69

4 Summary **71**

Appendix **74**

A 2D-Image Sets **75**
    A.1 J-images, TC-, Slope-DLIT, and MF-ILIT Images 75

Bibliography **81**

Abbreviations **89**

Acknowledgment **91**

Preface

Against the background of global warming due to greenhouse gases, the demand of new energy sources which reduce the emission of such gases rises steadily. Solar cells are one of the most promising canditates among alternative energy sources for generating electric energy with reduced emission of greenhouse gases. Driven by subsidies solar cell industry grew by about thirty per cent per year in the last decade world wide. In the course of this developement in solar cell industry also the demand on solar cell research increases. Research on new solar cell concepts as well as research on new materials are the main trends. Since solar cells are still a costly energy source, the main goal of all efforts made in solar cell research is to reduce the price of solar cells. This goal can be reached by reducing the costs for the used material, by increasing the efficiency of solar cells, and by increasing the yield of solar cell production.

About ninety per cent of the world's solar cell production are solar cells made of crystalline silicon. More than half of them are made from multicrystalline silicon. This material is of great interest for solar cell industry and solar cell research because the energy consumption for the production of multicrystalline silicon is much less and the production is therewith cheaper than that of monocrystalline silicon. The efficiencies reached by solar cells made of multicrystalline silicon solar cells are less than that of solar cells made from monocrystalline silicon due to the lower material quality of the multicrystalline material.

The efficiency losses of multicrystalline solar cells are caused by a lot of different reasons. One of the main tasks of solar cell research is to find out the physical origins of these efficiency losses and try to find ways to avoid them and therewith increase the yield of good solar cells in solar cell production. For that purpose solar cells are characterized by their current–voltage characteristics, which contain a lot of information about the electrical properties of a solar cell. The forward current–voltage characteristics of solar cells have been investigated intensively in the past. However, not all mechanisms which are decreasing the solar cell efficiency are understood in detail yet. Furthermore, the physical origins of the behavior of solar cells under reverse bias are only poorly investigated. Since solar cells may be damaged if they can not stand high reverse biases, they can not be used in modules. Therefore, a certain percentage of produced cells has to be discarded, which is an economical loss for the producers. Hence, there is a high interest to find out the physical origins of the unexpected reverse current–voltage characteristics of multicrystalline silicon solar cells. Nowadays, new materials as upgraded metallurgical grade silicon are used for silicon solar cell production. It will be shown that all problems leading to efficiency losses in solar cells made from conventional multicrystalline silicon will be enhanced in solar cells made from upgraded metallurgical silicon, because this material contains more harmful contaminations than conventional material. In particular, the breakdown voltage of such solar cells will be decreased and the number of unusable solar cells of such material will increase. Hence, the results of this work are very important also for the new silicon materials.

The established theoretical description of current–voltage characteristics of solar cells can not explain all features which can be found in current–voltage characteristics of real solar cells. For example, classical diode theory predicts a relatively low recombination current with an ideality factor of 2 or below. In reality, however, much higher recombination currents with an ideality factor of 3 and above are regularly found. Also some details of the physical origin of material induced ohmic shunts have been poorly understood in the past. In particular, the reverse current–voltage characteristics of real solar cells can not be described by the established two-diode model for p-n junctions. The reason for that is a lack of knowledge of the physical origins of breakdown in multicrystalline solar cells.

In the present work detailed experimental investigations on the reverse current–voltage characteristics of multicrystalline silicon solar cells are shown. These investigations reveal the nature of so-called "pre-breakdown" mechanisms in solar cells and explain the resulting experimental reverse current–voltage characteristics of solar cells. New measurement methods, which are especially developed for investigating the parameters of reverse biased solar cells are explained in detail. Furthermore, the present work gives an overview of effects influencing the forward bias $I-V$ characteristics of multicrystalline silicon solar cells, which affect their efficiency. Therefore, the work includes also a review of previous results in this field of our and other research groups.

The structure of this thesis is as follows: In the first chapter a brief description of the investigated type of solar cell is given, followed by an overview of the established theory of current–voltage characteristics of solar cells and their restrictions describing current–voltage characteristics of real solar cells, especially of reverse current–voltage characteristics. In the second chapter the established lock-in thermography imaging methods for revealing the currents flowing in solar cells are briefly described. The new lock-in thermography methods for investigating the reverse currents in solar cells, which were developed in this work, are demonstrated in detail. The first part of the third chapter deals with effects in forward current–voltage characteristics, where especially section 3.1.4 (Non-Linear Shunts) relies on earlier work of our group. The second part is an elaborate description of new experimental finding of the origins of features in reverse current–voltage characteristics of solar cells. The thesis is completed by a summary in chapter four.

Chapter 1
Solar Cells

Solar cells are large area p-n junctions and convert the energy of light, i.e. the energy of photons, into electrical energy. An introduction to the function of solar cells as well as solar cell concepts can be found in text books like [1, 2, 3].

The quality of p-n junctions is measureable by their current–voltage (I–V) characteristics. Therefore, measuring the I–V characteristics of solar cells is important to determine their quality. In this chapter, after a brief description of the investigated type of solar cells in section 1.1, a description of the established theory of I–V characteristics of solar cells is given in section 1.2. This is followed in section 1.3 by a description of established pre-breakdown mechansisms. In section 1.4 it is shown that typical real I–V characteristics of solar cells strongly deviate from these theoretical predictions. The explanation of these deviations is the topic of this work. Since all aspects of a solar cell, beginning from the feedstock used for the production, have an impact on the I–V characteristic of a solar cell, at first a description of the type of solar cell investigated in this work will be given.

1.1 Production of Multicrystalline Silicon Solar Cells

Since some design features, the material, and the manufacturing process of Si solar cells are important for some aspects of the behaviour of the I–V characteristics, a brief description of the material, the processing, and the design of solar cells is given in this part. Here, the manufacturing of commercial Si solar cells made from block-cast multicrystalline (mc) Si is described, since this type of solar cells is investigated in this thesis.

For multicrystalline Si solar cells typically solar-grade Si feedstock is used. Due to the high demand of Si nowadays also the use of upgraded metallurgical grade (UMG) Si becomes important in spite of the lower material quality. In [4] a short description of the different types of feedstock used for Si solar cells is given. Irrespective of the type of feedstock the production of mc-Si for solar cells is in principle the same.

Block-Casting Process of Silicon

The starting point of solar cell production are multicrystalline Si ingots. These ingots are produced by the block-casting process, which is based on the works of Bridgman [5] and Stockbarger [6] and is described in detail in [4, 7, 8]. Briefly, the silicon feedstock is melted in a quartz crucible, whose walls are covered with silicon nitride (Si_3N_4). Since most mc-Si solar cells are based on p-type wafers, the mc-Si ingots are p-doped by adding boron (B) to the feedstock befor melting in the crucible. The doped, liquid Si is cooled in a way that the Si crystallizes from the bottom to the top of the crucible. The grain boundaries of the solid mc-Si are columnar orientated, i.e. perpendicular to the future wafer surfaces (see Figure 1.3).

State-of-the-art solar cells are made from Si ingots with cuboidal shape having a mass of some hundred kilograms. Some cm of the sides, the top, and the bottom of the ingots are cut off, because this Si is highly polluted with foreign atoms reducing the solar cell performance. Afterwards the ingots are cut in bricks with a quadratic base of 156×156 mm². Finally, these bricks are cut into wafers of approximately 200 μm thickness[1]. Due to the boron doping the wafer have typically a resistivity of approximately 1 Ωcm, i.e. an acceptor dopant concentration of $N_A \approx 1.5 \times 10^{16}$ cm^{-3}.

Processing of Solar Cells

The basic processing steps of mc-Si solar cells are shown in Figure 1.1 and are described below. In Figure 1.2 a photo of the front side of an acidic texturized mc-Si cell with its gridlines and busbars is shown. A scheme of the typical layer structure of a commercial multicrystalline Si solar cell is shown in Figure 1.3.

1. The wafer surface is treated by an etch solution to remove the surface damage, which was caused by the wafer sawing process. The etching process increases the mechanical stability of the wafer, because micro-cracks and scratches are smoothed or etched away. In the same step the surface of the wafer is texturized. Texturization means that the flat surface of the wafer is roughened by

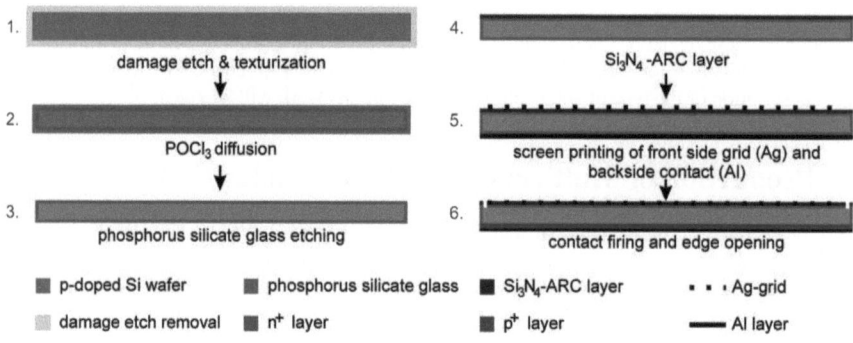

Figure 1.1: Processing steps of a crystalline Si solar cell, the steps are described in the text.

means of etching to improve the probability of absorption of light in the solar cell. Two different texturization techniques are widely used for texturization of industrial mc-Si solar cells [9]. Alkaline (KOH) etching, which leads to random pyramids on the Si surface due to anisotropically etching, and acidic (HF/HNO$_3$) etching, which is an isotropic etch and leads to scalloped surfaces.

2. An n$^+$-doped layer, the emitter, is formed on the p-type wafer. The sheet resistivity of the emitter is approximately 50 Ω/□, i.e. a peak donor concentration up to $N_D \approx 1 \times 10^{21}$ cm^{-3} is generated. For n-type doping of Si phosphorus (P) is used. Phosphorus is diffused into the wafer surface during a high temperature step (approximately 900 °C) using gaseous phosphorus oxychloride (POCl$_3$). A very thin n$^+$-doped Si layer of approximately 250 nm thickness is formed. Because of the high temperature

---
[1] Next generation Si solar cells may have areas of 210×210 mm² with thickness < 200 μm.

1.1. PRODUCTION OF MULTICRYSTALLINE SILICON SOLAR CELLS

during the diffusion impurities are gettered from the base into the emitter and to the surfaces. This increases the lifetime τ of excess charge carriers in the base.

3. During the POCl$_3$ step a phosphorus silicate glass layer is formed on the surface of the wafer. This layer has to be removed from the surface by etching with hydrofluoric acid (HF).

4. In order to increase light absorption, the emitter, which is the illuminated side (the front side) of the solar cell, is covered by an antireflection coating (ARC). The ARC is an approximately 70 nm thick non-stochiometric silicon nitride (SiN) layer grown by plasma-enhanced chemical vapor deposition (PECVD). The SiN contains high amounts of hydrogen. During the firing (step 6) hydrogen diffuses into the Si material and passivates defects and therefore again increases the material quality.

5. The electrical contacts of the solar cells are made by a grid of screen-printed metal contacts. The front side consists of many thin silver lines, called grid lines (vertical lines in Figure 1.2), which collect the current from the emitter layer, and two (sometimes three) thicker silver lines, called busbars (horizontal lines in Figure 1.2), which collect the electrical current from the grid lines. The backside of the solar cell is covered by an Al layer, which is typically approximately 30 µm thick. Since Al is not solderable, the back contact contains two thick silver lines, which have the same size and positions like the bus bars on the front side (see Figure 1.3). Both contacts are realized by screen printing of an Ag and an Al paste, respectively.

Figure 1.2: Front view of an acidic texturized, 156×156 mm² large mc-Si solar cell.

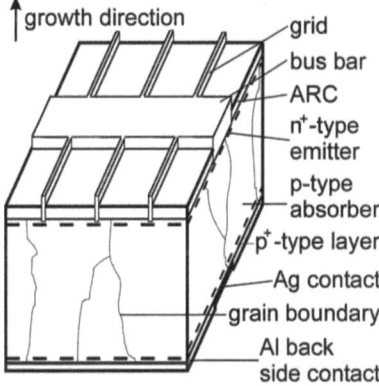

Figure 1.3: Layer structure of a typical industrial mc-Si screen printed solar cell.

6. By a subsequent short firing step at temperatures up to approximately 850 °C the Ag on the front side is etched through the SiN layer and forms an ohmic contact to the n$^+$-doped layer. The Al on the backside is an element of the third main group of the periodic system and therefore an acceptor in silicon. Due to the high temperature during the firing process Al atoms diffuse into the silicon and overcompensate the n-doping. Actually, during firing Al and Si are forming a liquid eutecticum layer, which recrystallizes in the cooling phase. Hence, an approximately 6 µm thick p$^+$-doped layer is formed, leading to a good ohmic contact on the backside of the cell. By the p$^+$-doped layer a so-called "back surface field" (BSF) is formed. The BSF reduces the recombination of electrons at the backside surface, since due to the high p-doping the concentration of electrons is smaller there and a reduction of the backside recombination velocity is achieved. After the firing of the contacts the edge of the solar cell is opened by a laser or by chemical or plasma etching, because the n$^+$-doped layer is

wrapped around the edge of the cell and therefore has direct contact to the p$^+$ BSF layer, which would lead to an unintentional short circuit at the edge of the solar cell.

1.2 Established Theory of Solar Cell I–V Characteristics

Solar cells are large area p-n junctions. Current-voltage characteristic measurements are one of the most common tools for characterizing p-n junctions. They reveal a lot of features of solar cells like leakage and shunt currents, recombination, and breakdown behavior, as well as series resistance effects. In this section the established theory of I–V characteristics of solar cells will be described. The differences between dark and illuminated I–V characteristics will be explained as well as the forward and reverse behavior.

I–V characteristics of solar cells consist of a forward and a reverse direction, shown in Figures 1.4 and 1.5. In forward direction the n$^+$-doped region of the p-n junction is connected to the negative pole and the p-doped region is connected to the positive pole. Therefore, the electrons in the n-region and the holes in the p-doped region flow towards the p-n junction and cross it easily, schematically shown in Figure 1.4b). In reverse direction the n-region is connected to the positive pole and the p-region is connected to the negative pole. The electrons and holes drift away from the p-n junction (see Figure 1.4a), therefore the charge carrier density in the vicinity of the p-n junction is close to zero (depletion region with width w, see Figure 1.4) and only a very low saturation current flows across the p-n junction.

1.2.1 One-Diode Model

I–V characteristics of p-n junctions (diodes) have been theoretically described by Shockley [10]. The relation between voltage and current of one non-illuminated diode reads as follows:

$$J_{\text{dark}} = J_{\text{S}}(e^{\frac{eU}{kT}} - 1). \tag{1.1}$$

Here J_{dark} is the current density[2] at a non-illuminated p-n junction in A/cm^2, J_{S} the saturation current density in this unit, e is the elementary charge, U the voltage applied to the junction, k the Boltzmann constant, and T the temperature. The saturation current density J_{S} is strongly affected by the recombination properties in the base region.

In forward direction, U is positive and the current density increases exponentially with voltage. In reverse direction the saturation current density J_{S} is reached already at very low voltages. The blue line in Figure 1.4 shows a plot of equation (1.1) in forward as well as in reverse direction. The diffusion current density described by equation (1.1) is indicated in the band diagram of Figure 1.4b) by a straight arrow symbolizing electron injection from the n$^+$- to the p-side of the junction.

If photons of adequate wavelength hits a solar cell, electron-hole pairs are generated. If the cell is connected to a load, a current can be extracted from the cell. This light-induced current density is called photocurrent density J_{photo} and is in very good approximation independent on U. It is symbolized in Figure 1.4 by the dashed-dotted arrows. To obtain the I–V characteristic of an ideal illuminated solar cell, equation (1.1) has to be expanded by J_{photo}. An illuminated solar cell is forward-biased, but the photocurrent is a reverse current, hence the photocurrent density J_{photo} must be subtracted from equation (1.1), which leads to

$$J_{\text{illu}} = J_{\text{S}}(e^{\frac{eU}{kT}} - 1) - J_{\text{photo}}. \tag{1.2}$$

---
[2]J is determined by dividing the overall current I of a solar cell by the cell area A.

1.2. ESTABLISHED THEORY OF SOLAR CELL I–V CHARACTERISTICS

If series resistances are neglected, the illuminated I–V characteristic described by equation (1.2) equals the dark I–V characteristic by equation (1.1) shifted by J_{photo}, which is schematically shown by the red dashed line in Figure 1.4. This so-called "superposition principle" holds in good approximation for silicon solar cells. At forward bias under illumination the dark current J_{dark} reduces J_{illu}. If an illuminated cell is operated under open circuit condition, which means no external current flows, J_{photo} equals J_{dark}. Thus I–V characteristics of illuminated cells are strongly affected by their dark I–V characteristics. Illuminated I–V characteristics define the important solar cell parameters, which

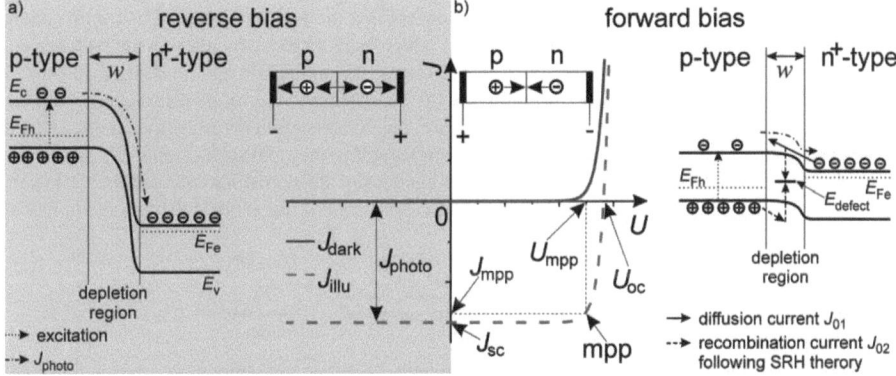

Figure 1.4: Ideal I–V characteristics of a solar cell: dark (solid curve) and illuminated (dashed curve); and the corresponding band diagrams of an illuminated solar cell for reverse a), and forward biasing b). Important solar cell parameters like the short circuit current J_{sc}, the open circuit voltage U_{oc}, and the mpp are noted in the graph. E_{defect} is the mid-gap level from the SRH theory, where recombination in the depletion region takes place. E_{Fe} and F_{Fh} are the quasi fermi levels of the electrons and holes, respectively. E_C and E_V are the conduction band and the valence band edges, respectively.

are shown in Figure 1.4. The short circuit current J_{sc}, which is influenced by recombination in the whole cell and the performance of the antireflection layer and the texturization, gives a number of how many photons are captured by the solar cell and are converted to a photocurrent. The open circuit voltage U_{oc}, which is strongly influenced by J_S, is a measure of the recombination in a solar cell and thereby of the material quality. The maximum power which can be extracted from a solar cell is the product of J_{mpp} and U_{mpp}. Here mpp stands for "maximum power point", it means that the rectangle formed by J_{mpp} and U_{mpp} (thin dashed lines in Fig. 1.4) has the largest area. This area, referred to the area formed by J_{sc} and U_{oc}, is called fill factor (FF):

$$\text{FF} = \frac{J_{mpp} U_{mpp}}{J_{sc} U_{oc}}. \tag{1.3}$$

The efficency η of a solar cell is by definition the quotient of the power density generated at mpp and the incident power density on the cell area:

$$\eta = \frac{J_{mpp} U_{mpp}}{P_{incident}} = \frac{\text{FF} J_{sc} U_{oc}}{P_{incident}}. \tag{1.4}$$

1.2.2 Two-Diode Model

The exponential relationship of Shockley's equation (1.1) is not sufficent to describe the I–V characteristics of solar cells. By equation (1.1) only the diffusion current (J_{01}) of the diode is described, shown by the solid arrow in the band diagram of Figure 1.4b). However, in p-n junctions besides the diffusion current also the recombination current (J_{02}) occurs, which is schematically shown by dashed arrows in the band diagram of Figure 1.4b). The electrons recombine with holes in the space charge region at a defect level with energy E_{defect}. By applying the recombination theory of Shockley, Read [11] and Hall [12] (SRH theory) to the recombination in the space charge region, a second diode describing the recombination current density J_{02} is defined, which considers recombination effects in the depletion region [13]. If $E_{\text{defect}} = \frac{E_{\text{gap}}}{2}$ (see Figure 1.4b) the recombination is most effective, which leads to an ideality factor n = 2 in equation (1.5). E_{gap} is the bandgap of the semiconductor.

Hence, the two-diode model comes from the concept that the p-n junction is acting as one diode describing the part of the diffusion current density, which leads to recombination in the bulk region, and a second diode describing the part of the space charge recombination current density, illustrated by the equivalent circuit of a solar cell in Figure 1.6. Furthermore, the series resistance R_s and the parallel or shunt resistance R_p, both given in in Ωcm^2, are considered in the two-diode model. These two parameters are very formally added here without regarding their physical origins. By adding the diffusion current density and the recombination current density and taking into account that R_s lowers the voltage at the p-n junction and that R_p decreases the current at the diodes, the two-diode model reads as follows:

$$J = \underbrace{J_{01}(e^{\frac{e(U-JR_s)}{kT}} - 1)}_{\text{diffusion}} + \underbrace{J_{02}(e^{\frac{e(U-JR_s)}{nkT}} - 1)}_{\text{recombination}} + \underbrace{\frac{U-JR_s}{R_p}}_{\text{resistance}}. \quad (1.5)$$

Here J_{01} is the diffusion saturation current density and J_{02} is the recombination saturation current density. As described above, the ideality factor n of the recombination current term is n = 2 in SRH theory for a midgap level. If the level is not at midgap, also ideality factors between 1 and 2 can be explained by SRH theory, but not ideality factors above 2 [14].

In Figure 1.5 a comparison between an I–V characteristic calculated by equation (1.1) and equation (1.5) is shown. Equation (1.1) (dashed red line in Figure 1.5) only describes the part of the I–V characteristics, which is dominated by the diffusion current, whereas equation (1.5) also describes the recombination current, which dominates the I–V curve at lower forward voltages. If there is a considerable parallel shunt resistance R_p (dashed-dotted blue line in Figure 1.5) it usually dominates the low-voltage part of the characteristic. The series resistance R_s reduces the slope at high currents. It will be shown in the next sections that equation (1.5) fits measured I–V curves much better than equation (1.1), but the recombination current and the reverse characteristic are wrongly described.

It should be noted here that, even for infinitely large R_p, also the two-diode model predicts a linear I–V characteristic for $U < kT$. If the two exponential terms in Figure 1.5 are expanded for small U

$$J = \left(\frac{J_{01}e}{kT} + \frac{J_{02}e}{nkT}\right)U \quad (1.6)$$

appears. This is equivalent to an

$$R_p^* = \frac{1}{\left(\frac{J_{01}e}{kT} + \frac{J_{02}e}{nkT}\right)} \quad (1.7)$$

without the exponential terms. Thus, it is clearly wrong to measure R_p just by evaluating the linear part of the characteristic for $eU < kT$, as it is often being done. Instead, the whole forward characteristic has to be fitted to equation (1.5).

1.2. ESTABLISHED THEORY OF SOLAR CELL I–V CHARACTERISTICS

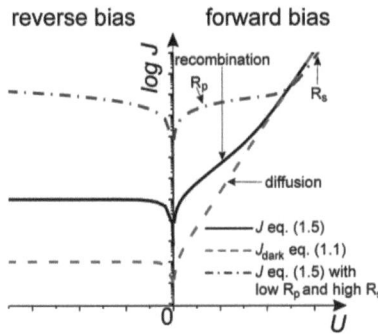

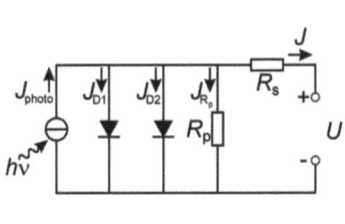

Figure 1.5: $I–V$ curves calculated by the two-diode model given by equation (1.5) with infinite R_p (solid curve) and finite R_p (dashed-dotted curve). The dashed curve shows the curve calculated by Shockley's equation (1.1).

Figure 1.6: Equivalent circuit of a solar cell with serial resistance R_s, shunt resistance R_p, and the two diodes representing the diffusion current density and the recombination current density.

Reverse-Biased Solar Cells

According to this theory, in the absence of ohmic shunts, the current of a reverse-biased ideal p-n junction is determined by the diffusion saturation current density J_S from equation (1.1), which is reached already at very low reverse voltages and is only some fA/cm². J_S is made up only of those charge carriers which are generated thermally as minority charge carriers in the corresponding region and which are not recombined before reaching the n-region for electrons and the p-region for holes, respectively. Hence only those minority charge carriers, which are generated at a distance of approximately one diffusion length away from the correspondig majority charge carrier region, contribute to J_S, which is expressed by:

$$J_S = \left(\frac{eD_e n_i^2}{L_e N_A} + \frac{eD_h n_i^2}{L_h N_D} \right). \tag{1.8}$$

with D_e and D_h being the diffusivities of electrons and holes, n_i the intrinsic charge carrier density, L_e and L_h the diffusions lengths of electrons and holes, and N_A and N_D the acceptor dopant and donor dopant concentration, respectively.

Regarding the two-diode model the saturation current density of solar cells is the sum of the diffusion and recombination saturation current densities J_{01} and J_{02}, which should be in the order of nA/cm². In Figure 1.5, in the absence of ohmic shunts, the reverse $I–V$ curves saturate accordingly to equation (1.8), dashed curve, and equation (1.5), solid curve. However, it will be shown in the next sections that in real solar cells the reverse behavior of $I–V$ curves can neither be fitted by the one-diode model nor by the two-diode model. Real reverse $I–V$ characteristics and especially the causes of pre-breakdown in solar cells will be discussed in this thesis in more detail later on. The next section will start with a description of the established theory of breakdown in p-n junctions.

1.3 Breakdown Mechanisms in p-n Junctions

Regarding equations (1.1) or (1.5) the reverse current of a p-n junction should be limited by the saturation current densities. However, at high reverse voltages a steep increase of the reverse current is observed, which is called breakdown. The physics of breakdown in p-n junctions was already reviewed exhaustively in the 1950s and 1960s, and an overview was given by Mahadevan [15]. The breakdown models discussed there are based on nearly defect-free monocrystalline semiconductor material and do not consider the high density of defects in multicrystalline material, which is used for mc-Si solar cells.

1.3.1 Avalanche Breakdown

The mechanism of avalanche breakdown (AB) is schematically shown in Figure 1.7. Electrons in the space charge region, which are accelerated by the electric field of a reverse-biased p-n junction, may gain enough energy to generate an electron-hole pair by impact ionization. If the newly generated electrons and holes gain also enough energy to generate further electron-hole pairs, a charge carrier avalanche will be produced. The number of electrons in the conduction band increases quickly and results in a high current flow across the p-n junction. Then breakdown due to avalanche occurs. Avalanche breakdown (AB) is also called impact ionization breakdown. It is characterized firstly by a negative temperature coefficient (TC) of the breakdown current and secondly by the multiplication of charge carriers. The negative TC is due to electron energy loss caused by increased phonon scattering at higher temperatures. Therefore, the probability of electron-hole pair generation decreases with increasing temperature.

The multiplication of charge carriers can be measured by illuminating a reverse-biased p-n junction. Then the multiplication is expressed by the ratio of the photocurrent at a certain reverse voltage $I(U)$ under avalanche conditions to the constant photocurrent at a low reverse voltage $I(U_{low})$, at which no avalanche occurs. This ratio is the avalanche multiplication factor MF:

$$\text{MF}(U) = \frac{I(U)}{I(U_{low})} \qquad (1.9)$$

For a flat p-n junction with a typical base doping concentration of solar cells of $N \approx 1 \times 10^{16}$ cm^{-3}, the avalanche breakdown voltage should be about -60 V [16]. A more detailed description of AB and its properties can be found in [17].

1.3.2 Zener Effect

The second breakdown mechanism is the Zener effect [18], schematically shown in Figure 1.8. The basic mechanism of the Zener effect is tunneling of an electron from the valence band to the conduction band. The probability of tunneling depends on the field strength at the p-n junction. For significant currents due to band-to-band tunneling an electric field of $\mathscr{E} \geq 10^6 \frac{V}{cm}$ is needed in Si [19]. In Si p-n junctions such fields are achieved if the base doping concentration is $N \geq 5 \times 10^{17}$cm^{-3} [20]. Under these conditions the Zener breakdown voltage in Si is $U_b \approx -5$ V [19, 20]. Note that at a doping concentration of $N \approx 5 \times 10^{17}$cm^{-3} in Si, avalanche multiplication becomes the dominant breakdown mechanism already at a voltage of $U \approx -7$ V [19]. Hence, Zener breakdown may only occur in very narrow highly doped p-n junctions [21] at low reverse voltages. Since electrons are emitted from the valence band to the conduction band, the Zener effect is also called internal field emission (IFE). The

1.4. REAL I–V CHARACTERISTICS

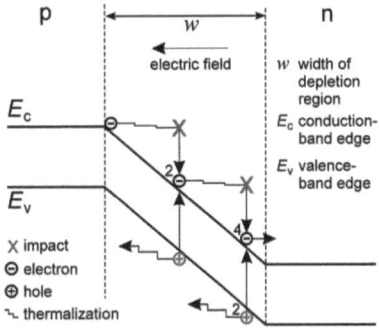

Figure 1.7: Avalanche effect schematically. The figures state the number of electrons and holes after each impact ionization.

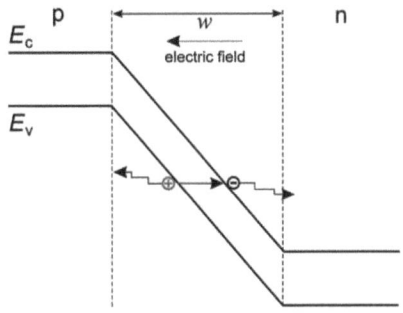

Figure 1.8: Internal field emission (Zener effect) schematically. The symbols are explained in the legend of Figure 1.7.

current due to Zener effect may be high enough to produce breakdown in p-n junctions [22]. The tunneling current J_t is given by [19]:

$$J_t = \frac{\sqrt{2m^*}e^3 \mathscr{E} U}{4\pi^2 \hbar^2 \sqrt{E_{gap}}} \exp\left(-\frac{4\sqrt{2m^*}E_{gap}^{3/2}}{3e\mathscr{E}\hbar}\right). \quad (1.10)$$

Here m^* is the effective mass of electrons in Si, U the reverse voltage, $\mathscr{E}$ the electrical field, which also depends on U by $\mathscr{E} \propto \frac{U}{w}$, and $\hbar$ the reduced Planck constant.

Zener breakdown is characterized by a positive temperature coefficent of the current, because the bandgap becomes smaller at higher temperatures and thus tunneling becomes more probable. For Zener breakdown no carrier multiplication is expected. Since the base doping concentration in solar cells is only 10^{16} cm^{-3}, Zener breakdown should be improbable here.

1.4 Real I–V Characteristics

In Figures 1.9 and 1.11 a measured forward and reverse I–V characteristic of a typical industrial screen printed, acidic texturized mc-Si solar cell without any ohmic shunts is shown by the open circles. The physical quantities affecting real I–V characteristics are described in this section.

1.4.1 Forward I–V Characteristics

In Figure 1.9a) different effects on the shape of a real forward I–V characteristic are shown. The short-dashed blue curve in Figure 1.9a) represents the diffusion current part of the I–V characteristic. The intersecion of this line with the current density axis is giving the value for J_{01}. The dashed red curve in Figure 1.9a) represents the part of the I–V characteristic which is caused by the recombination current. The intersection of that line with the current density axis gives the value for J_{02}. A low parallel (shunt) resistance R_p of the solar cell would further increase the current at low voltages, which is denoted in Figure 1.9a) by the green dotted-dashed curve. The serial resistances R_s flattens the I–V characteristic at higher voltages, because the serial resistance becomes important at higher currents and linearizes the characteristics. This effect is shown in Figure 1.9a) by the orange double-dotted-dashed line. The measured curve (open circles) was fitted by the two-diode model and the fit is shown by the solid line in Figure 1.9a).

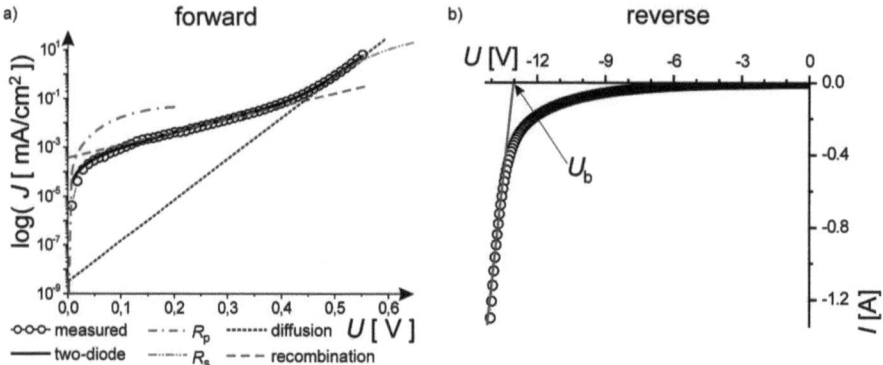

Figure 1.9: Measured forward a) and reverse b) I–V characteristic of a standard industrial solar cell (open circles) with negligible high R_p. The influences on the shape of the forward I–V curve are schematically indicated in a) (explained in the text). In b) the reverse characteristics of the solar cell plotted linear for the estimation of U_b is shown.

Fit Parameters for Two-Diode Model

The parameters for the fit by equation (1.5) are extracted from the measured I–V characteristic. From the intersections of the extrapolated diffusion current density curve (short-dotted curve in Figure 1.9a) and the extrapolated recombination current density curve (dashed curve in Figure 1.9a) the values of J_{01} and J_{02} for the fit are taken: $J_{01} = 3 \times 10^{-9}$ $\frac{mA}{cm^2}$ and $J_{02} = 4 \times 10^{-4}$ $\frac{mA}{cm^2}$. The value of J_{01} is at least in the correct order of the value expected by theory for a lifetime of 10 μs, but J_{02} is 2 to 3 orders of magnitudes higher than expected. McIntosh [14] has calculated J_{02} depending on the lifetime τ of the charge carriers and the resistivity of the Si material.

For the typical dopant concentration of solar cells of $N \approx 3 \times 10^{16}$ cm^{-3} McIntosh calculates J_{02} to:

$$J_{02} = 10^{-12} \frac{mA}{cm^2} \frac{1 s}{\tau[s]}, \tag{1.11}$$

which would lead to $J_{02} = 10^{-7} \frac{mA}{cm^2}$ for the measured solar cell here, assuming an electron lifetime of $\tau = 10$ μs.

To achieve a good fit with the measurement n differs significantly from the theoretical value 2. Here n is found to be 3.2. Since the cell measured for Figure 1.9 was a cell without shunts (high R_p) and no serial resistance problems (low R_s), typical values of $R_p = 1 \times 10^4$ Ωcm^2 and $R_s = 1 \times 10^{-4}$ Ωcm^2 were used for the fit.

Regarding n=3.2, which is significantly higher than the expected value of 2 and the unexpectedly large value of J_{02}, even the two-diode model assuming SRH theory seems to be not sufficient for describing all physical mechanims of real forward I–V characteristics.

1.4.2 Previous Models for the Recombination Current

Queisser already has noticed the unexpectedly large ideality factor of solar cells and has suspected its cause in metallic precipitates in the material [23]. However, such precipitates are not present in typical solar cells.

Kaminski et al. have attributed this current to trap-assisted tunneling or field-assisted recombination at point defect levels [24]. Since for increasing forward bias the electric field in the depletion region reduces, trap-assisted tunneling becomes less probable for increasing forward bias, which is equivalent to a saturation effect leading to an increase of the ideality factor. However, the concentration of really existing point defects in solar cells is much too small to account for the really measured magnitude of the recombination current.

Schenk and Krumbein have tried to solve this problem by assuming two-level (pair) recombination with participation of at least one shallow level [25]. Due to their low energy, and since shallow levels have a spatially more extended orbital, trap-assisted tunneling is much more probable for shallow levels than for deep levels. However, since single shallow levels show a very weak coupling to the more distant band, Schenk had to assume infinitely strong coupling between the shallow level and a second deep one or a complimentary shallow level. In reality this coupling should be exponentially dependent on the distance of the levels to each other, so assuming infinitely strong coupling is a very coarse assumption. Nevertheless, also Schenk was not able to explain the magnitude of measured recombination currents quantitatively based on the concentration of known defect levels.

Breitenstein and Heydenreich and later on also many other authors have published the observation that the recombination current is not proportional to the area but to the circumference length of a cell [26]. Hence it should be essentially due to recombination at surface states in the position where the p-n junction crosses the surface. This edge recombination current density has to be given in units of A/cm (per cm edge line) instead of A/cm^2 as was usual previously.

Kühn et al. performed realistic device simulations of the edge region and found that, if the edge region is treated only as a recombining surface without any surface charging (according to SRH recombination theory), the finite thermal velocity of carriers limits the recombination current density to values below 10^{-8} A/cm^2, which is much below typically measured values [27]. Kühn et al. showed that this limitation may be overcome by assuming oxide-induced surface charges leading to a depleted surface. This depletion occurs also naturally at any extended defects in doped semiconductors with a high density of states due to capture of majority carriers. The corresponding band bending locally expands the recombination region and thus allows to explain large surface recombination current densities. This effect was already considered by Breitenstein and Heydenreich, but the theoretical description proposed there was wrong, since it did not regard any surface recombination. The effect of a surface depletion region on edge recombination is schematically shown in Figure 1.10, which is based on [26]. Since the n$^+$-emitter is highly doped, only depletion of the p-surface is assumed here. The barrier at the surface leads to a "groove" (see Figure 1.10b) where electrons from the highly doped emitter are injected into. The recombination with holes occurs then in an extended region of the p-surface shown by the thick red line in Figure 1.10b), which is the final reason enabling higher recombination currents. Whereas, at the p-n junction only directly at the p-n junction recombination occurs as shown by the short thick red line in Figure 1.10a). McIntosh treated the edge current only by SRH theory, leading to an ideality factor of 2 [14]. However, he explicitely regards the series resistance in the emitter layer on the way to the edge region. As it was mentioned already in sections 1.2 and 1.4, this resistance indeed tends to increase the ideality factor. However, for a single diode it actually only creates a "hump" in the I-V-characteristic, rather than a large ideality factor over an extended bias range. For fitting such characteristics McIntosh had to assume rather unrealistic combinations

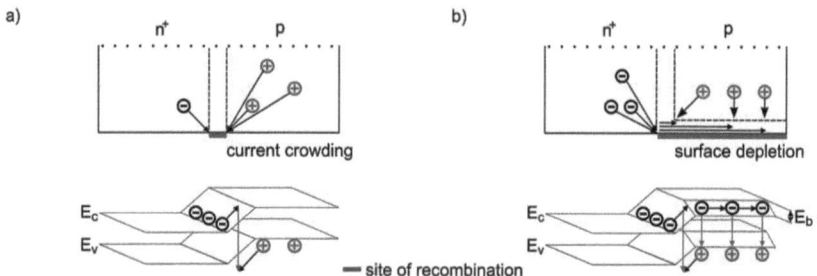

Figure 1.10: In a) the recombination at the p-n junction is shown, b) shows the additional recombination at the "groove", E_b is the barrier hight.

of very different series resistances to different edge regions. Especially this theory cannot explain large ideality factors at low forward biases below 0.2 V, since at such low voltages the recombination current is so small that the series resistance does not play any role. It will be shown below that indeed the series resistance increases the ideality factor of the recombination current, but it is not the primary reason for a large ideality factor. There must be intrinsic reasons leading to a large ideality factor for recombination at extended defects like surfaces.

Hence, in spite of many new results, a satisfying physical model for describing the recombination current in solar cells was still missing so far. New findings on this topic will be introduced in section 3.1.4.

1.4.3 Reverse $I-V$ Characteristics

The reverse direction of the measured $I-V$ characteristics is shown in linear representation in Figure 1.9b). It shows a low and moderate slope up to a certain reverse voltage. From that voltage on a steep increase of the current occurs, this voltage is named breakdown voltage U_b. By extrapolating the steepest part of the curve to $I = 0$ the breakdown voltage U_b can be estimated. This leads to $U_b \approx -13$ V for the measured $I-V$ curve in Figure 1.9b).

Conventionally on the y-axis of the $I-V$ plots current densities of solar cells are indicated, assuming a homogenously current flow through the cell area. For forward currents, at least for the photocurrent and the diffusion current in monocrystalline cells, this approximation applies. For the shunt and recombination currents, especially for the reverse current, this approximation does not apply for solar cells, because most of these currents flows at localized sites (see below). As a consequence, for the forward characteristic the current density J is plotted, whereas for the reverse characteristic the current I is plotted (e.g. in Figure 1.9b). Here the cell area is 243 cm^2. Note that the measured reverse current is orders of magnitudes higher then predicted by SRH theory and does not show the expected saturation behavior. The breakdown voltage of -13 V is significantly lower than the -60 V expected for avalanche breakdown at this doping concentration.

Breakdown Regions

Figure 1.11 shows semi-logarithmically the plot of the measured reverse $I-V$ characteristic of the same solar cell as in Figure 1.9. The reverse $I-V$ characteristic is divided into three regions, namely linear, defect-induced pre-breakdown and hard pre-breakdown, which are schematically shown and

1.4. REAL I–V CHARACTERISTICS

separated by thin vertikal solid lines in Figure 1.11. Close to the origin the reverse characteristic shows a linear behavior and the current at -2 V is already much higher than expected from equation (1.5). In the defect-induced pre-breakdown region the reverse current increases moderately superlinearly and is also significantly higher than expected from equation (1.5). In the hard pre-breakdown region the reverse current increases strongly as shown in the example in Figure 1.11 from $U_b \approx -13$ V on. A lot definitions for U_b can be found in the literature, which are summarized e.g. in [15]. Since several physical mechanisms may be responsible for pre-breakdown, there is no unambiguous definition for U_b. Mostly the point in the reverse I–V characteristic, at which the slope of the reverse current increases considerably compared to the slope of the rest of the I–V characteristic is used to define U_b. However, the definition of U_b is generally arbitrarily since breakdown in p-n junctions is not a threshold process.

Reverse Models

The two-diode model given in equation (1.5) considers none of the above described reverse behaviors. In Figure 1.11 the two-diode model (solid curve) and the measured reverse I–V characteristic (open circles) are compared. Obviously equation (1.5) does not fit the measurement. Other approaches to fit reverse I–V characteristics of solar cells by different authors are given in the following. In Figure

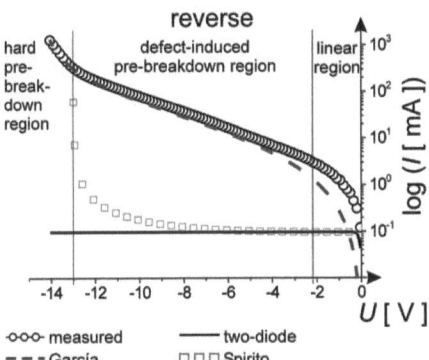

Figure 1.11: Measured reverse I–V characteristic of a standard industrial solar cell (open circles). The linear, the pre-breakdown and the hard pre-breakdown region of the reverse characteristic are shown as well as the fits of the two-diode model (solid line) and the models of Spirito (rectangles) and García (dashed line).

1.11 fits by different models are shown. One of the first attempts to study reverse characteristics was made by Rauschenbach [28] in 1972 (not shown here). Later different authors proposed models to fit the reverse characteristics of solar cell by means of avalanche breakdown [29, 30, 31]. A nice overview about these models is given by Alonso-García [32], who contributes another model in that publication.

All these models are empirical and are based more or less on Bishop [29], whose one-diode model was adapted to the two-diode model by Quaschning et al. [33]. Here two of the reverse models reviewed in [32] will be exemplarily shown. These models are used to fit the measured reverse characteristic and are shown in Figure 1.11. For the fit with Spirito's model [30] a breakdown voltage of $U_b = -13$ V and a dark current of $I_{dark} = 9.7 \times 10^{-2}$ mA from the measured I–V curve, which equals the recombination saturation current $J_{02} = 4 \times 10^{-4}\ \frac{mA}{cm^2}$ (see Figure 1.9a), are used. J_{01} is neglected, because it is orders of magnitudes smaller than J_{02}. The models by Spirito [30] and Hartman [31] are equal, and exhibit a poor correlation with the measurement. In both models the avalanche breakdown is considered by a multiplication coefficient MC:

$$I = \left(I_{sc} - I_{dark}(e^{\frac{qU}{n_{01}kT}} - 1)\right) MC(U), \tag{1.12}$$

with

$$MC(U) = \frac{1}{1 - (U/U_b)^m}, \tag{1.13}$$

and $m = 3$ the Miller exponent [34], $I_{sc} = 0$ A, because we are regarding dark I–V characteristics (I_{sc} is the short circuit current). Lopez Pineda (equation (4) in [32]) extended Spirito's equation by the influence of a shunt resistance R_p, which does not change the reverse I–V characteristic significantly and therefore is not shown here. None of the models mentioned here fits the real reverse characteristic.

The model of García (equation (6) in [32]) considers the saturation currents in the term I_N and fits the reverse I–V characteristic quite well, which is shown in Figure 1.11. García's model reads as follows:

$$I = \frac{I_N}{1 - K_e} = \frac{I_{sc} - G_p U + cU^2}{1 - e^{\left(B_e\left(1 - \sqrt{(U_d - U_b)/(U_d - U)}\right)\right)}} \tag{1.14}$$

with K_e the multiplication coefficient, B_e a dimensionless constant (described in [35]), G_p the shunt conductance, c a coefficient (described in [32]) and U_d the built-in voltage with a value of approximately 0.9 V. For the fit $I_{sc} = 0$ mA, $B_e = 3$, $G_p = 1 \times 10^{-4}\ \frac{1}{\Omega}$, $c = 0.27\ \frac{1}{\Omega^2 mA}$ and a breakdown voltage of $U_b = -15.35$ V are used. U_b is higher than examined from Figure 1.9b), because García uses a different definition for determining the breakdown voltage [32]. Ruíz and García proposed to weight the different currents in the numerator of equation (1.14) [35], but the physical origins of the weighting factors respectively the microstructural breakdown mechanism are still unclear. The fit-parameters of the model of García are empirical, and in none of the presented models any defect states contributing to the reverse behavior of solar cells are discussed.

Only Bishop and Simo tried to find the physical reasons of the formation of hard avalanche breakdowns (hot-spots) in solar cells. Bishop mentioned small pits in the surface of the solar cell causing hot-spots [36], and Simo found an increased densitiy of defects in the solar cell material at hot-spot sites [37]. However, he was only able to speculate about the physical mechanism of the hot-spot formation. In particular the underlying physical mechanism causing the breakdown effect was still unclear.

Chapter 2

Measurement Methods

In this chapter measurement methods are introduced which are helpful to image the distribution of currents in solar cells. A very powerful tool for this purpose is lock-in thermography (LIT) and its different modes, which are described in section 2.1. In particular, in section 2.1.1, new LIT methods for imaging physical parameters of pre-breakdown sites in solar cells are presented, which were developed in this thesis. These techniques are very important for investigating the behavior of reverse-biased solar cells and permit detailed analysis of the lateral distribution of different physical mechanisms.

Afterwards, in section 2.2, brief introductions are given into the use of electroluminescence (EL) and in section 2.3 of electron microscopy methods like scanning electron microscopy (SEM), electron beam induced current (EBIC), and transmission electron microscopy (TEM) for solar cell research. The latter methods are very important to study the microscopic structures in solar cells influencing the behavior of $I-V$ characteristics. In section 2.4 different light microscopy techniques are described and some details of electrical measurements are given.

Finally a short overview of special methods to image and characterize precipitates in multicrystalline Si, which may cause shunts in solar cells, is given.

2.1 Lock-in Thermography

Lock-in thermography is a camera based imaging method to determine small temperature deviations in the device under test. In solar cells temperature deviations may be caused by heating due to higher local current densities at defects compared to the expected homogeneous current density. The higher the current density at a certain position the higher is the temperature at this position. Since in solar cells the temperature deviations are in the range of some mK down to some µK, the lock-in technique has to be used to enhance the signal-to-noise ratio.

Here the basics of LIT are described, for a detailed description of LIT techniques the reader is refered to the textbook by Breitenstein and Langenkamp [38].

The Lock-in Principle

Lock-in is a technique to separate small periodic signals from statistical noise. For that purpose it is necessary to pulse the primary signal periodically with the so-called lock-in frequency $f_{\text{lock-in}}$ before the signal is amplified and detected. The oscillating alternating current (a.c.) part of the signal is converted into a direct current (d.c.) signal by rectifying the signal. Controlled by a reference trigger, the negative part of the signal is converted into a positive signal, whereas the positive part of the signal passes the rectifier directly without any change. The noise, which is also rectified, is suppressed by a small-band filter, whose center frequency equals the signal frequency. Now only the small-band noise contributes to the output signal. This remaining noise is statistical noise and averages out to a value of zero if the output signal is integrated over a certain time. Only the direct current signal is left. With lock-in technique it is possible to separate signals, which are orders of magnitudes smaller than the surrounding noise [38].

The lock-in output signal S is described by the multiplication of the detected signal $F(t)$ with a symmetric correlation function $K(t)$. For the integration over a certain integration time t_{int}, S is [38]:

$$S = \frac{1}{t_{int}} \int_0^{t_{int}} F(t)K(t)dt \text{ with } K(t) = \begin{cases} +1 & \text{(first half period)} \\ -1 & \text{(second half period)} \end{cases} \qquad (2.1)$$

The correlation function $K(t)$ has to be symmetric so that after a number of complete lock-in periods within the integration time the integral over $K(t)$ is exactly zero. For LIT digital lock-in correlation is used. Therefore, $F(t)$ and $K(t)$ are transformed into numbers of the measured values F_k and the weighting factors K_k, respectively, t_{int} as to be substituted by the number of measurements M, so S becomes a sum [38]:

$$S = \frac{1}{M} \sum_{k=1}^{M} F_k K_k \qquad (2.2)$$

For several reasons it is advantageous to perform LIT with digitization events, which are syncronized to $f_{lock-in}$. For solar cells the synchronization of the lock-in frequency with the digitization events is easy to manage [38]. Therefore, in one lock-in period a certain number h of digitization events takes place, with $h \geq 4$ [38]. The measurement is averaged over H lock-in periods and in each of these lock-in periods the weighting factors K are the same. The digital lock-in correlation reads then as follows [38]:

$$S = \frac{1}{hH} \sum_{i=1}^{H} \sum_{j=1}^{h} K_j F_{i,j} \qquad (2.3)$$

In our LIT system the two-channel correlation, i.e. two sets of correlation functions, approximating a sine and cosine, are used. The advantage of the two channel correlation is that the phase and amplitude informations of the signal are considered. The first channel, measures the signal in phase with the sine, this is the $S^{0°}$ signal. The second channel measures the signal in phase with the cosine, i.e. 90° phase-shifted to the first channel, called $S^{90°}$ signal. Since in lock-in thermography the $S^{90°}$ signal is negative, the –cosine is used, i.e. the $S^{-90°}$ signal, to have both signals positive. The weighting factors then read as follows [38]:

$$K_j^{0°} = 2\sin\left(\frac{2\pi(j-1)}{h}\right), K_j^{-90°} = -2\cos\left(\frac{2\pi(j-1)}{h}\right) \qquad (2.4)$$

and the amplitude A and phase Φ of the signal are [38]:

$$A = \sqrt{(S^{0°})^2 + (S^{-90°})^2} \qquad (2.5)$$

$$\Phi = \arctan\left(\frac{-S^{-90°}}{S^{0°}}\right) \text{ (−180° if } S^{0°} \text{ is positive).} \qquad (2.6)$$

2.1. LOCK-IN THERMOGRAPHY

Lock-in Thermography at Solar Cells

In Figure 2.1 a camera based lock-in thermography measurement setup is shown schematically. Our system is a commercial lock-in system *TDL 640-XL* from Thermosensorik GmbH Erlangen [39]. It is equipped with a camera containing an indium-antimonide (InSb) detector array with 640×512 pixels, which is sensitive for wavelengths from 1 to 5 μm (near- to mid-infrared).

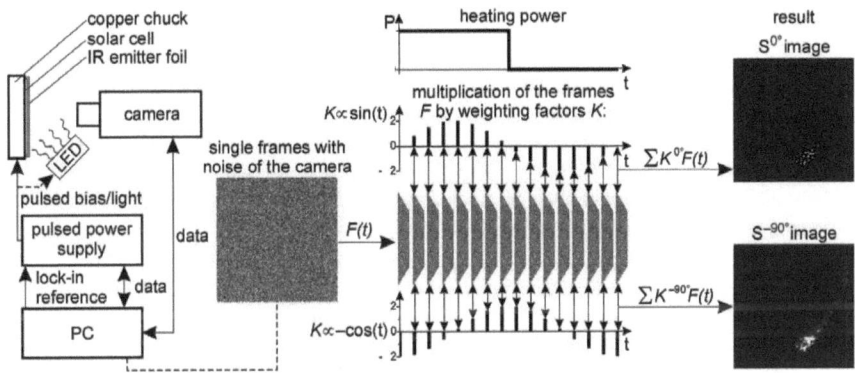

Figure 2.1: Scheme of a camera based lock-in thermography imaging setup for solar cells. The solar cell is pressed on a copper chuck by an infrared (IR) emitter foil, which is sucked on the chuck by a vacuum pump. The principle of the sine/-cosine correlation is shown schematically for $h = 15$ frames per lock-in period. The results shown here are the $S^{0°}$ and $S^{-90°}$ images of a measurement of a solar cell. For LED-illumination (ILIT) a transparent foil has to be used. The Figure is adapted from [38].

For lock-in thermography it is necessary to have a periodic temperature signal from the solar cell. For dark lock-in thermography (DLIT) this is provided by applying a square-shaped pulsed voltage U_{pulsed} to the solar cell. If the solar cell is illuminated with pulsed or continuous light, the technique is called illuminated lock-in thermography (ILIT). The illumination is indicated by the light emitting diodes (LED) in Figure 2.1. The voltage or light pulses are defined by the lock-in frequency $f_{\text{lock-in}}$ [38]

$$f_{\text{lock-in}} \leq \frac{f_s}{h}, \qquad (2.7)$$

which is provided by the personal computer (PC). f_s is the image sampling rate of the camera in Hz, and h is the number of frames per lock-in period. For h the condition $h \geq 4$ holds [38]. The lock-in correlation procedure is then applied to each pixel of each image, which is schematically shown in Figure 2.1. From the $S^{0°}$ and $S^{-90°}$ images the phase-independent amplitude image A and the phase image Φ are calculated by equations (2.5) and (2.6), respectively.

The lock-in procedure provides four images, the measured 0° and –90° images and the calculted amplitude and phase images. The 0°, –90° images and the amplitude image are scaled in mK, the phase image is scaled in degree. Additionally to these four images a topography image, which allows to compare the positions of the heated sites with the topology of the solar cell, is taken by the camera. The image quality, i.e. the signal-to-noise ratio of a lock-in thermography image, is, among other parameters, mostly influenced by the lock-in frequency $f_{\text{lock-in}}$ and the number of lock-in periods H (measurement time). These influences and others are discussed in detail in [38]. Here only some basic rules are highlighted.

The image from a localized heat site is blurred, because the thermal wave propagates in all directions into the material. The heat conductivity λ of Si is very high (148 $\frac{W}{m \cdot K}$). Therefore, heat signals in Si solar cells are broadened strongly and the spatial resolution of heat sources by LIT is limited. This is expressed by the so-called thermal diffusion length Λ [38]:

$$\Lambda = \sqrt{\frac{2\lambda}{\rho c_p 2\pi f_{\text{lock-in}}}}, \qquad (2.8)$$

with λ the heat conductivity, ρ the density, and c_p the specific heat of the material. Λ depends on the material, and on the lock-in frequency by $1/\sqrt{f_{\text{lock-in}}}$, which means the higher the lock-in frequency the better is the spatial resolution of heat sources in a LIT image. For example, in Si Λ is 3 mm at a lock-in frequency of 3 Hz. However, the higher $f_{\text{lock-in}}$ the smaller is the signal and therefore the measurement time has to be increased in order to achieve a good signal-to-noise ratio.

The signal-to-noise ratio is improved by increasing the measurement time t_{meas}, i.e. the number of lock-in periods. The average amplitude of the noise for constant $f_{\text{lock-in}}$ behaves like [38]: $\langle A_{\text{noise}} \rangle \propto \frac{1}{\sqrt{t_{\text{meas}}}}$.

The $S^{0°}$ image of a LIT measurement reveals the highest spatial resolution of heat sources in solar cells [38]. This can be seen from the comparision of the $S^{0°}$ image with the $S^{-90°}$ image in Figure 2.1. It has to be noted here that the solar cells investigated in this work can be regarded as thermally thin samples, which means that the thermal diffusion length Λ is large compared to the thickness of the solar cell. Therefore, the $-90°$ phase component of the LIT measurements has to be used for any quantitative analysis [38]. The $S^{-90°}$ image is proportional to the locally dissipated power and therefore can be used for techniques which are described in detail in [38, 40].

2.1.1 Imaging of Physical Parameters of Forward Biased Solar Cells

Forward $I-V$ characteristics of solar cells are influenced by the shunt resistance R_p, the serial resistance R_s and the diffusion- and recombination-current densities. All these parameters can be imaged by dark lock-in thermography (DLIT) and illuminated lock-in thermography (ILIT).

Imaging of Shunts, Recombination- and Diffusion Current by DLIT

Shunts are represented either by very low R_p at localized sites, or by certain areas in solar cells where for any other reason a locally increased current flows inside the cell. At these specific sites the solar cell heats up if a voltage is applied to the cell or if the cell is illuminated. A detailed description of different types of shunts in crystalline Si solar cells is given partly later in this thesis and in [41].

Shunts are divided into two types: linear (ohmic) shunts and non-linear shunts. For shunt imaging by DLIT a pulsed voltage U_{pulsed} is applied to the solar cell and the distinction between linear and non-linear shunts can be easily made by taking a DLIT image at $U_{\text{pulsed}} = 0.5$ V forward bias and $U_{\text{pulsed}} = -0.5$ V reverse bias, respectively. In Figure 2.2 typical DLIT images of a solar cell containing linear and non-linear shunts are shown. If the amplitude signal A at a shunt position depicts the same strength in both images the shunt is a linear one, as shown in Figure 2.2a) and b) by the dotted arrow. On the other hand, if A is different in both images the shunt is a non-linear shunt. This is shown in Figure 2.2a) and b) by the solid arrows.

If $U_{\text{pulsed}} \leq 0.5$ V, primarily localized currents like ohmic shunt currents and recombination currents are imaged by DLIT (cf. Figure 1.9a), which is shown in Figure 2.2a). The reasons for non-linear

2.1. LOCK-IN THERMOGRAPHY

shunts are manifold, mostly they may be caused by enhanced recombination current at scratches, cracks, or other extended defects (see section 3.1.4). Especially the DLIT signals at the opened edges of the solar cell are caused by enhanced recombination current as can be seen in Figure 2.2a) at the dashed arrow. At voltages $U_{\text{pulsed}} > 0.5$ V, the 2-dimensional diffusion saturation current density can be imaged, this is shown in Figure 2.2c). This high current DLIT-image anti-correlates with the lifetime image (cf. Figure 2.9b), because the regions of high diffusion current correspond generally to regions of low lifetime.

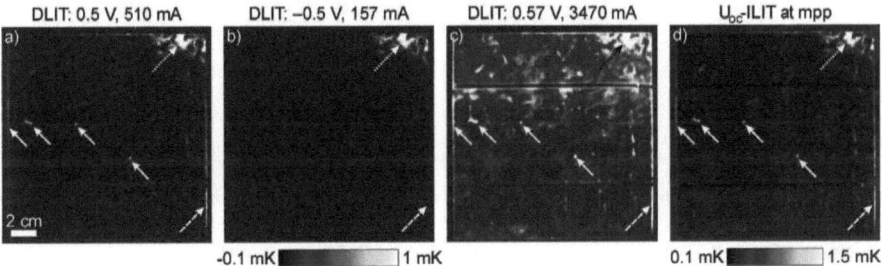

Figure 2.2: Amplitude DLIT images of a solar cell: a) at forward bias, b) at reverse bias, and c) at higher forward bias of 0.57 V, here the 2-dimensional signal shows the distribution of the diffusion saturation current density J_{01}. In d) a U_{oc}-ILIT image of the cell at mpp $U = 0.5$ V is shown.

Imaging of Shunts by ILIT

ILIT was invented independently by Isenberg [42] and Kaes [43] in the year 2004. To image shunts with ILIT the solar cell is illuminated homogenously with pulsed light of ideally 850 nm to 950 nm wavelength. If the solar cell is not contacted, i.e. no current flows in an external circuit, this technique is called U_{oc}-ILIT. Depending on the illumination intensity it is possible to image the cell at different points on the illuminated $I-V$ characteristics, e.g. at mpp or at U_{oc} (cf. Figure 1.4). The advantages of U_{oc}-ILIT over DLIT is that the solar cell needs no contacts and that non-linear shunts, which are caused by recombination sites, are a bit stronger pronounced than in DLIT. The disadvantage is that it is impossible to distiguish between linear and non-linear shunts, since the cell under open circuit is always forward biased.

In Figure 2.2d) an U_{oc}-ILIT image of the solar cell is shown. The image was taken at mpp, i.e. 0.5 V. The ILIT signals at the non-linear shunts are a bit stronger than in the forward DLIT image in Figure 2.2a). Please note that the U_{oc}-ILIT image is scaled differently than the DLIT images in Figure 2.2. Of course, also the linear shunts are clearly visible in the U_{oc}-ILIT image as can be seen in Figure 2.2d) at the dotted arrow.

Imaging of Serial Resistance

The serial resistance R_s of solar cells can be imaged by DLIT under high forward bias or by the R_s-ILIT technique [44]. For imaging R_s by DLIT a high forward bias is applied to the cell, because R_s has only a resonable effect at high currents. Then regions of high R_s appear dark, because no or less diffusion current flows there. The disadvantage of this technique is that it only works well for monocrystalline Si solar cells, where homogenous material quality and therefore homogenous

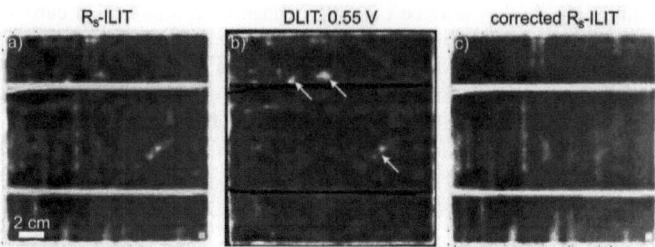

Figure 2.3: a) R_s-ILIT image, b) DLIT image measured at 0.55 V showing shunts, and c) R_s-ILIT image corrected by the DLIT image.

lifetimes can be assumed. Multicrystalline Si solar cells show an inhomogenous lifetime distribution. Therefore, it is hard to distinguish between regions of high lifetime (which appear also dark in the high-current DLIT image) and regions of high R_s. It had to be proposed to solve this problem by yielding the ratio of a high-current and a medium-current DLIT image [44]. However, this technique did not become popular yet since high-lifetime regions appear very noisy.

R_s-ILIT overcomes the problems of R_s-DLIT by performing the ILIT measurement under continous illumination and applying a pulsed bias to the solar cell. Due to the illumination the ILIT signal is generally higher than the DLIT signal. The bias is pulsed between short circuit (0 V) and mpp (0.5 V). At regions of high R_s the ILIT-singal becomes less negative or positive, this is explained in detail in [44]. Drawbacks of R_s-ILIT are firstly that the image can not be interpreted quantitatively, since the signals depend non-linearly on R_s, and secondly that the R_s-ILIT image also shows shunts. The latter drawback can be corrected by substracting a shunt image, taken by DLIT ideally at 0.55 V, from the R_s image [44]. Figure 2.3 shows an R_s-ILIT image, the corresponding DLIT shunt image at 0.55 V and the shunt-corrected R_s-ILIT image. A technique to evaluate the serial resistance quantitively is the *REcombination current and Series resistance Imaging* (RESI) technique by Ramspeck et al. [45]. This technique is a combination of lock-in thermography and electroluminescence imaging.

Imaging of the Ideality Factor n

With DLIT it is also possible to map the ideality factor n [40]:

$$n(x,y) = \frac{e(U_2 - U_1)}{kT \ln\left(\frac{S_2^{-90°}(x,y)U_1}{S_1^{-90°}(x,y)U_2}\right)}, \quad (2.9)$$

where U_1 and U_2 are forward voltages in a small range near the maximum power point, for example $U_1 = 0.5$ V and $U_2 = 0.55$ V are appropriate values. $S_1^{-90°}$, $S_2^{-90°}$ are the corresponding $-90°$ images. In Figure 2.4 DLIT images at 0.5 V and -0.5 V and the n-factor map of an mc-Si solar cell are shown. The shunts, which are marked by arrows in Figure 2.4, and the edges of the solar cell show an enhanced n-factor of up to $n \geq 4$.

2.1. LOCK-IN THERMOGRAPHY

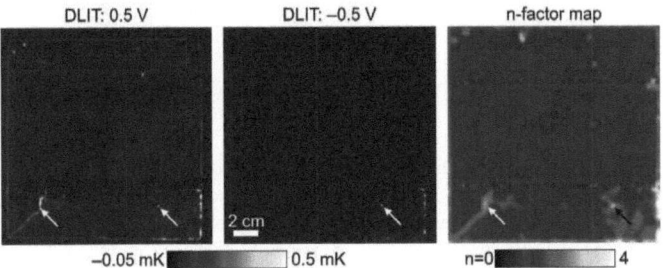

Figure 2.4: DLIT images at different voltages and map of the ideality factor n of an mc-Si solar cell.

2.1.2 Imaging of Physical Parameters of Reverse-Biased Solar Cells

The LIT techniques presented so far are techniques to image physical parameters of forward-biased solar cells. In the last years the interest on the reverse $I-V$ characterisic and pre-breakdown behavior of solar cells increased. In this section new LIT methods are presented, which were especially developed in this work for imaging physical parameters of pre-breakdown sites in solar cells. These new methods are published in [46].

There are two main breakdown mechanisms, namely internal field effect (IFE) and avalanche breakdown (AB), which have already been described in section 1.4. The main physical parameter to distinguish the two breakdown mechanisms is the temperature depedence of the reverse current. AB shows a strong negative temperature coefficient, whereas IFE shows a slightly positive TC, regarding the reverse current, respectively. Furthermore, AB is characterized by the multiplication of charge carriers. In the following section new LIT techniques to image the temperature coefficent of the reverse current, the slope of the reverse current, and the multiplication factor of the avalanche breakdown (see equation (1.9) are described.

DLIT Current Density Images

To image the temperature coefficient and the slope of the current at reversed biased solar cells, standard DLIT measurements have to be performed. At first the measured DLIT images, which are scaled in mK, have to be converted into current density images scaled in $\frac{mA}{cm^2}$. For this purpose it is necessary to use the $S^{-90°}$ image, because in thermally thin samples like solar cells only these images are proportional to the locally dissipated power density [40].

To convert an $S^{-90°}$ image into a current density image the average temperature signal $\langle T \rangle$ of the $S^{-90°}$ image has to be calculated. If (x,y) is the size of the $-90°$ DLIT image in pixels, $\langle T \rangle$ reads as follows:

$$\langle T \rangle = \frac{\sum_{x=0}^{x}\sum_{y=0}^{y}S^{-90°}(x,y)[\text{mK}]}{x \cdot y}. \tag{2.10}$$

If the imaged area is A with the corresponding total current of that area I, and the temperature signal at each pixel $S^{-90°}(x,y)$ in mK, then the current density $J(x,y)$ at each pixel is:

$$J(x,y) = S^{-90°}(x,y)\frac{I}{\langle T \rangle A}. \tag{2.11}$$

In the following the term (x,y) will be skipped for simplification, so that $J(x,y) = J$ and $S^{-90°}(x,y) = S^{-90°}$. Note that this procedure works only accurately if the cell has no series resistance problems and if the flowing currents are moderate.

Temperature Coefficent DLIT

For imaging the temperature coefficient, the solar cell is imaged at different temperatures T_j. This is realized by tempering the copper chuck, and thereby the attached solar cell (see Figure 2.1), by a water thermostat to the desired temperature with an accuracy of $\Delta T = \pm 0.5$ °C. At each temperature several DLIT images of the solar cell at various reverse voltages U_i are taken. It is important that all DLIT images are taken at the same position of the cell in the image. This assures that each pixel of an image matches the corresponding pixel in the other images. All DLIT images, measured at T_j and U_i, have to be converted into current density images $J_{i,j}$ by applying the procedure in equations (2.10) and (2.11). From the $J_{i,j}$ images a two-dimensional set of the current density images at different temperatures and reverse voltages can be generated, which is very helpful to analyze the results.

The temperature coefficient (TC) image is calculated by substracting two current density images $J_{i,j}$ and $J_{i,j-1}$ taken at the same reverse voltage U_i, but at consecutive temperatures T_j and T_{j-1}. This difference is normalized to the average of $J_{j,i}$ and $J_{j,i-1}$ and the temperature difference. The resulting image is called TC-DLIT, which is calculated as follows:

$$\text{TC-DLIT} = \frac{2(J_{i,j} - J_{i,j-1})}{(T_j - T_{j-1})(J_{i,j} + J_{i,j-1})} \quad (2.12)$$

The TC-DLIT image shows the distribution of the TC in the reverse-biased solar cell and is scaled, if equation 2.12 is multiplied by 100, in %/K. Figure 2.5 shows an example of a TC-DLIT image of an mc-Si solar cell. The TC-DLIT image is independent of the magnitude of the reverse current density.

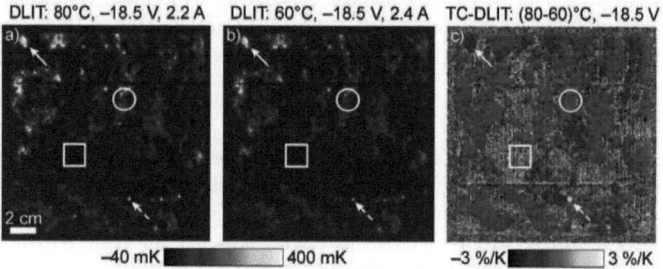

Figure 2.5: −90° DLIT images of an mc-Si solar cell measured at a) $U_i = -18.5$ V and $T_j = 80$ °C and b) $U_i = -18.5$ V and $T_{j-1} = 60$ °C. In c) the TC-DLIT image calculated from the images in a) and b) is shown. All regions appearing dark indicate a negative TC, the regions appearing bright indicate a positive TC of the local pre-breakdown sites. In particular: The solid arrow marks a pre-breakdown site of clearly negative TC, the circle mark a pre-breakdown site of TC about 0 %/K, the square marks a region which seems to be pure noise, but is dominated by positive TC (as will be shown later). The dashed arrow points at a site where clearly positve TC is measured.

Equation (2.12) is normalized to the average current density $(J_j + J_{j-1})/2$ and therefore describes the slope of the current density J at the midpoint of the two temperatures $T_{\text{mid}} = (T_j + T_{j-1})/2$. Mathematically this is a central-difference derivate at T_{mid}, which gives the correct value for linear functions. For non-linear functions the difference $T_j - T_{j-1}$ has to be suitably small to get correct values, according to the fact that the central-difference derivate is a difference quotient.

Slope-DLIT

To image the slope of the local reverse current at a constant temperature T_j, DLIT images at two biases U_i and U_{i-1} are taken. Analogously to TC-DLIT imaging these −90° DLIT images are converted into current density images by applying equations (2.10) and (2.11). These images are then used to calculate the so-called slope-DLIT image by subtracting the current density image at U_{i-1} from that at U_i and normalizing this difference to the voltage difference and the average of the current density images leading to an equation similar to equation (2.12):

$$\text{slope-DLIT} = \frac{2(J_{i,j} - J_{i-1,j})}{(U_i - U_{i-1})(J_{i,j} + J_{i-1,j})}. \tag{2.13}$$

Slope-DLIT images show the slope of the current in reverse-biased solar cells independently of their magnitude and are scaled in V^{-1}, or, if equation (2.13) is multiplied by 100, in %/V. With slope-DLIT images it is easy to distinguish between sites of different steepness of the current in solar cells, which is examplarily shown in Figure 2.6.

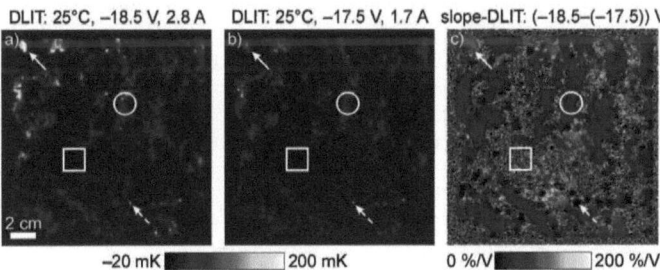

Figure 2.6: −90° DLIT images of an mc-Si solar cell measured at a) $U_i = -18.5$ V and $T_j = 25$ °C and b) $U_{i-1} = -17.5$ V and $T_j = 25$ °C. In c) the slope-DLIT image calculated from the images in a) and b) is shown. All regions appearing bright indicate a steep slope of the current, the regions appearing dark indicate a flat slope of the current. Since for this example the same solar cell like in Figure 2.5 was used the symbols mark the same positions in that solar cell. The solid arrow marks a pre-breakdown site of a high slope, the circle marks a pre-breakdown site of a moderate slope. At the position of the dashed arrow a spot of very low slope can be seen. The signal in the square is very noisy since it is outside of pre-breakdown regions, and no specific slope can be determined here.

Multiplication Factor ILIT

With multiplication factor ILIT (MF-ILIT) it is possible to image the local multiplication of charge carriers due to the avalanche effect. Therefore, MF-ILIT is the method of choice to distinguish between regions of avalanche breakdown and regions of Zener breakdown in solar cells. The multiplication factor, see equation (1.9), is the ratio of the photocurrent density at a given reverse voltage $J(U)$ and the photocurrent density $J(U_{\text{low}})$ at a voltage U_{low}, where evidently no breakdown occurs.

For MF-ILIT the solar cell is biased at a constant reverse voltage and is illuminated by pulsed light of 850 nm wavelength with an intensity of about 0.1 sun. The light excites electron-hole pairs in the solar cell, which leads to a photocurrent. This photocurrent is homogenous for a good solar cell, and at such low illumination intensity the cell heats up homogenously mainly due to thermalization of the charge carriers. Hence, the $S^{-90°}$ ILIT image at a low reverse voltage, where no pre-breakdown in the solar cells occurs, is homogenous. At higher reverse bias, local pre-breakdown may occur. If this pre-breakdown is not influenced by the photocurrent, it should remain invisible in the lock-in image, since the reverse bias is constant. However, if the pre-breakdown is due to avalanche, additional charge carriers are generated and the current density at these positions will be increased periodically, leading to an increased heating and an inhomogenous $S^{-90°}$ ILIT image. For quantitative analysis, one has to divide the current image, taken at the higher voltage, by the current image, taken at U_{low}, which is the definition of the multiplication factor (see equation (1.9).

Since the current density in an illuminated solar cell is the sum of the generated photocurrent and the dark current, the images can not be converted into current density images as described for the DLIT techniques in this section. The $S^{-90°}$ images are proportional to the dissipated power density p. Therefore, the modulated current density can be described by dividing the $S^{-90°}$ signal image by the overall relaxation voltage U_{relax} [47]:

$$J = \frac{p}{U_{relax}} \propto \frac{S^{-90°}}{U_{relax}}. \tag{2.14}$$

Here $S^{-90°}$ is the local temperature modulation in mK. The overall relaxation voltage is the sum of the applied reverse voltage U, the diffusion voltage of the p-n junction U_D, which is the barrier hight at zero voltage, and the thermalization voltage $U_{thermal}$, which is schematically shown in Figure 2.7a). U_D for p-n junctions of typical mc-Si solar cells with a base doping of 3×10^{16} cm^{-3} is approximately 0.95 V. $U_{thermal}$ depends on the wavelength of the irridiated light and is in our case, for Si with $E_{gap} = 1.13$ eV at room temperature, and $\lambda = 850$ nm:

$$U_{thermal} = \frac{\frac{hc}{\lambda} - E_{gap}}{e} = 0.33 \text{ V} \tag{2.15}$$

with c the light velocity. Then the current density is proportional to

$$J(U) \propto \frac{S^{-90°}(U)[\text{mK}]}{U + U_D + U_{thermal}}, \tag{2.16}$$

this procedure has to be applied to both ILIT images taken at U_{low} and U, which leads, according to equation (1.9), to the calculation procedure for the MF-ILIT images:

$$\text{MF-ILIT} = \frac{(U_{low} + U_D + U_{thermal})S^{-90°}(U)}{(U + U_D + U_{thermal})S^{-90°}(U_{low})}. \tag{2.17}$$

In Figure 2.7 an example of an MF-ILIT image of the same cell which was already shown in Figures 2.5 and 2.6 (TC-DLIT and slope-DLIT) is presented. In Figure 2.7a) the determination of the overall voltage U_{relax} is schematically shown by means of a band diagram. In Figure 2.7b) and c) the $S^{-90°}$ ILIT images at a low reverse voltage and at a high reverse voltage are shown, respectively. Figure 2.7d) shows the MF-ILIT image calculated from b) and c).

General Aspects of the TC-, Slope-DLIT and MF-ILIT Imaging Methods

In Figure 2.8 the TC-DLIT image, the slope-DLIT image and the MF-ILIT image of the cell which was already shown in Figures 2.5, 2.6, and 2.7 are shown. That comparision gives a nice overview of the values of the different physical parameters of the pre-breakdowns distributed in the solar cell. To analyze the trends of the physical parameters of pre-breakdown sites in a solar cell it is very helpful to

2.1. LOCK-IN THERMOGRAPHY

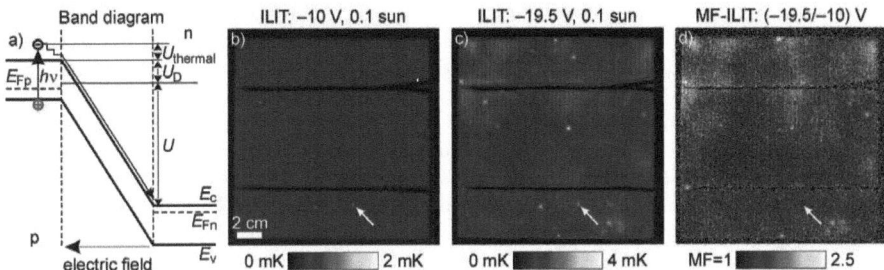

Figure 2.7: a) determination of the relaxation voltage for equation (2.17). b) $S^{-90°}$ image at -10 V reverse voltage, the image is homogenous except for some bright spots, one of them is marked by the arrow. c) $S^{-90°}$ image at -19.5 V reverse voltage, the image is much more inhomogenous then b), the new bright areas are regions of avalanche pre-breakdown. In d) the MF-ILIT image is shown. The scaling denotes how much electrons each "original" electron was able to ionize. At the arrow no MF is measurable, maybe this signal comes from a weak ohmic shunt, so here no avalanche occurs.

create 2-dimensional arrays of the images, for example by displaying the TC-DLIT images versus the voltage, which leads to an array of TC-DLIT$_{U_i, T_j - T_{j-1}}$. This holds also for slope-DLIT images, where the slope-DLIT$_{U_i - U_{i-1}, T_j}$ are displayed, as well as for MF-ILIT images, if the MF-ILIT procedure is done for different temperatures. Examples for such kinds of 2-dimensional arrays are given in Figures A.1, A.2, A.3, and A.4 in the appendix.

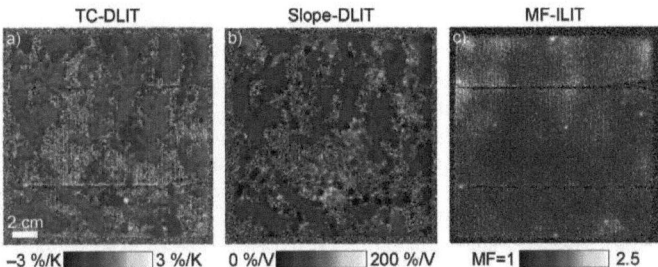

Figure 2.8: Comparision of the three new imaging methods for investigating pre-breakdown sites in solar cells.

Due to the lateral spreading of the heat in the Si material it is not possible to determine the exact positions of the pre-breakdown sites with TC-DLIT and slope-DLIT. Here the values of the pre-breakdown sites appear not as point like sources but as extended areas of the same signal strength. That is due to the fact that the whole region being thermally influenced by the shunt displays the correct result, because of the normalization performed in equation (2.12) and equation (2.13). For strong pre-breakdown sites this region may extend over many thermal diffusion lengths Λ.

The extended areas of the same values of TC or slope images are due to the broadening of the thermal signal of the point-like pre-breakdown sites and the normalization to the average current density. The shape of the signal of a point-like source measured by LIT can be mathematically described by a point spread function, whose spatial extension is determined by the thermal diffusion length Λ. Since the same point spread function is contained in all signals in the DLIT images used to calculate the TC- and slope-mappings, the spatial resolution of the TC- and slope-DLIT images gets lost due to the normalization, whereby only the constant ratio of the current strengths is obtained. Therefore, the TC and slope signals of the point-like pre-breakdown sites appear 2-dimensionally broadened and are bounded by the noisy regions where the normalization does not lead to a constant value.

For the determination of the exact position of pre-breakdown sites, the original DLIT images or other methods like electroluminescence and scanning electron microscopy techniques have to be used. These techniques are described briefly in the next sections. For TC-DLIT as well as for slope-DLIT it is important to know that for both imaging methods the signal can get into saturation. Assuming that the smaller of the two current density values is negligible against the larger one, equations (2.12) and (2.13) reduce to:

$$\text{TC-DLIT} = \frac{2}{(T_j - T_{j-1})} \text{ and slope-DLIT} = \frac{2}{(U_i - U_{i-1})}. \quad (2.18)$$

The TC-DLIT and slope-DLIT values saturate to the these values, because at this specific point the current density, which is used for normalization, increases proportional to the absolute slope of the current density. Therefore the relative slope is constant according to (2.18). However, this is only critical for the measurements of slope-DLIT, because the current behaves strongly non-linear with the voltages U_i. To overcome this drawback, the difference $U_i - U_{i-1}$ should be decreased as soon as the measured value comes close to that of equation (2.18).

2.2 Luminesence Methods

In the last years electroluminescence (EL) and photoluminescence (PL) imaging methods became versatile tools for solar cell characterization. The advantages, in particular for EL, are the comparatively inexpensive and simple measurement setup and the high spatial resolution, which can be achieved with EL/PL methods. Here only a brief introduction of the measurement setups and the results, which can be obtained by luminescence methods, is given.

In 2005 Fuyuki et al. showed the determination of the minority carrier diffusion length in polycrystalline Si solar cells by EL [48]. Recently a number of different EL methods were developed. Trupke et al. showed PL imaging of the minority carier lifetimes [49]. Breitenstein et al. [50] and Kasemann et al. [51] discussed the possibilities of shunt detection by EL imaging. Ramspeck et al. [45], Hinken et al. [52], and Kampwerth et al. [53] showed EL-based methods for determining the serial resistance in solar cells.

For EL the Si solar cell is simply biased to a forward or a reverse voltage and the emitted light is detected by a charge coupled device (CCD) camera. Under forward bias, the emitted light comes from the recombination of electrons, which are injected from the emitter into the base, where they are minority carriers. This is called band-to-band electroluminescence. Recombination at defect levels in the bandgap can also take place, and the emitted light is then called defect band emission [54].

Luminescence light at reverse-biased Si p-n junctions was observed by Hall, Burch, and Newman [55]. The origin of the luminescence light is still under discussion. Figielski and Toruń claimed luminescence at reverse-biased Si p-n junction is attributed to bremsstrahlung of hot carriers in a Coulomb field of charged carriers [56]. Yamada and Kitao published that this luminescence is rather due to recombination radiation [57]. The discussion about the origin of the luminescence light at reverse-biased Si p-n junction is out of the scope of this thesis. Here, the light emission at reverse-biased solar cells is used only to determine the sites of pre-breakdown with a better spatial resolution than it is possible with LIT methods. While the spectrum of forward-bias EL is centered at about 1100 nm, reverse-bias EL shows a broad spectrum extending into the visible region.

2.2. LUMINESENCE METHODS

In PL the excitation of electrons is done by illuminating the Si solar cell with light of adequate wavelength, e.g. 850 nm. The detection of the emitted light can be done with the same camera as for EL, because the same radiative recombination mechanisms as at EL occur. However, special filtering is needed, because the excitation light has to be suppressed. At forward bias the band-to-band EL/PL signal S_{lum} depends strongly on the local voltage U_{local} in the base material, which is the difference of the two quasi Fermi levels there:

$$S_{lum} = Ce^{\frac{eU_{local}}{kT}}, \qquad (2.19)$$

where C is a calibration constant describing the optical and material properties [58]. According to Fuyuki C is proportional to the local diffusion length [48]. Since the band-to-band luminescence in Si solar cells has a wavelength corresponding to the bandgap of Si ($E_{gap} = 1.13$ eV leading to $\lambda = 1100$ nm), it is sufficient to use a comparatively inexpensive Si-CCD camera for the detection of the luminescent light. Though the quantum efficiency of a Si-CCD device at 1100 nm is only some percent, it is sufficient to cool the CCD chip thermo-electrically to about −20 °C and integrate the EL/PL signal over some seconds to get resonable results.

In Figure 2.9 a DLIT image taken at 0.6 V and EL images taken at a forward voltage of 0.6 V and at −16 V reverse voltage of an mc-Si solar cell are shown. The signal in the DLIT image anti-correlates to the signal in the forward bias EL image. The increased signal in the DLIT image is due heating by a locally increased diffusion current. The dark areas in the forward bias EL image in Figure 2.9b), which almost matches the bright areas of the DLIT image in Figure 2.9a), are regions of locally reduced voltage thus leading to a lower EL signal correspondig to equation (2.19). One reason for the decreased local voltage is e.g. increased non-radiative recombination at defects in the bulk. Therefore, dark regions in EL images are regions of lower minority charge carrier lifetime. The bright spots in the reverse-bias EL image are due to light emission from pre-breakdown sites. It is obvious that the regions of pre-breakdown in that solar cell also match quite well to the regions of lower lifetime, a comprehensive analysis about that matter is given later in this work. Ohmic (linear) shunts are not leading to reverse-bias EL emission. However, it should be mentioned that at linear shunts the local voltage drops dramatically so that also a decrease in the EL intensity is the consequence. Therefore, it is not trivial to distinguish between regions of decreased charge carrier lifetime and particularly weak linear shunts in a solar cell by EL imaging [50, 51].

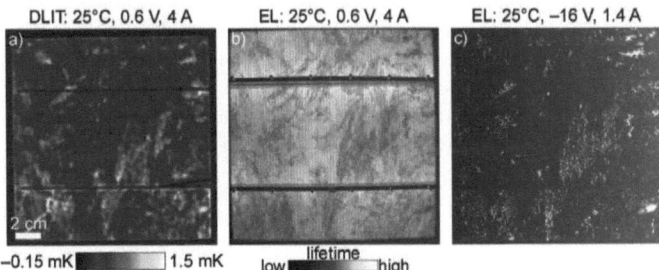

Figure 2.9: Comparision of a) a high current DLIT image, b) a forward EL image and c) a reverse EL image of one and the same cell.

2.3 Electron Microscopy Methods

The methods introduced so far are able to image the physical parameters influencing $I-V$ characteristics of solar cells on a macroscopic level. However, most of the underlying mechanisms have microscopic origins. In order to get a deeper insight into these mechanisms, the samples (solar cells), pre-characterized by LIT or EL/PL imaging, have to be analyzed by microscopic methods. The most powerful tools for that purpose are scanning and transmission electron microscopes. In this section no basic informations about SEM and TEM techniques are given, but some special methods are introduced, which are important for the analysis of solar cells and which are used in this work.

2.3.1 SEM Techniques

Scanning electron microscopy (SEM) is important to image surfaces of solar cells with a very high spatial resolution and is very important to clarify the origins of shunts. The strength of scanning electron microscopes are special techniques, which are able to image electrically active defects in solar cell material (see below) or to perform element analysis by energy-dispersive x-ray (EDX) analysis. EDX is used in this work to determine the element composition of precipitates, which may occur in mc-Si for solar cells.

Electron Beam-Induced Current Technique

Electron beam-induced current (EBIC) is used to image electrically active defects, the shape of p-n junctions, and also shunts in solar cells. A review about EBIC is given in [59]. The EBIC method is described briefly here. The electron beam of the SEM scans the sample, which is electrically connected to a current amplifier. The electron beam hits the sample and generates electron-hole pairs. Since the sample is connected in a circuit with the amplifier, a current is induced due to the electron beam, just as for light illumination. This current is amplified and represents the signal, which is imaged according to common SEM technique. If electrons recombine at defects, e.g. electrically active grain boundaries in the Si material, the EBIC current decreases, which is displayed in the EBIC image by a decreased EBIC signal. If the beam scans over a p-n junction or another kind of electronic barrier, where charge carriers are collected, an increased EBIC signal occurs.

In this work a special EBIC technique, called lock-in EBIC, described in [60], was used to image avalanche pre-breakdown sites in mc-Si solar cells. For imaging pre-breakdown sites it is necessary to apply a high d.c. reverse voltage to the solar cell sample. This d.c. signal would overload the EBIC amplifier. Therefore an electric circuit, see Figure 2.10 [61], which is similar to that introduced in [62], is used for the measurements to avoid d.c. currents at the EBIC amplifier. For lock-in EBIC the electron beam of the SEM is chopped to get a pulsed signal from the sample, and the EBIC signal is fed to a lock-in amplifier. Thereby signal distortions due to the R-C coupling are avoided. The advantage of lock-in EBIC over conventional EBIC is a much better signal-to-noise ratio in the presence of noisy pre-breakdown currents and the avoidance of a.c.-coupling artefacts.

2.3.2 TEM

Transmission electron microscopy is used to characterize crystal defects and structures of the size of some nanometer occuring in solar cell material. EDX in transmission electron microscopes and special element analysis methods like electron energy loss spectroscopy (EELS) are able to determine the element composition of very small clusters e.g. in grain boundaries of solar cells, which are

2.4. INFRARED MICROSCOPY & ELECTRICAL MEASUREMENTS

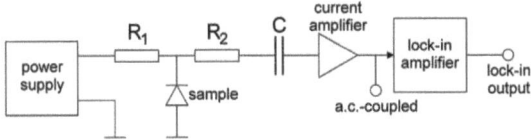

Figure 2.10: Electrical circuit to apply a reverse voltage to a solar cell sample for the analysis with lock-in EBIC.

decorated with foreign atoms. For TEM elaborate sample preparations, for instance focused ion beam (FIB) cuts, are necessary. In this work TEM in combination with FIB preparation of the samples was used to determine the crystallographic structure of precipitates in mc-Si and the microscopic origins of pre-breakdown sites in mc-Si solar cells. The TEM work was cordially performed by A. Lotnyk (now with Christian-Albrechts-Universiät zu Kiel, Germany) and N. Zakharov (MPI Halle, Germany). The FIB preparation was kindly done by H. Blumtritt (MPI Halle, Germany).

2.4 Infrared Microscopy & Electrical Measurements

Transmission Infrared Microscopy

Infrared microscopy (IRM) was used for the first time (to my knowledge) at Technische Universität Bergakademie Freiberg [63] to analyze precipitates found in mc-Si for solar cells [64, 65].

Infrared microscopy is an imaging technique exploiting that moderately doped Si becomes transparent for wavelengths of 1.1 µm to 10 µm. The Si sample is penetrated by the microscope light, but if precipitates exist in the Si bulk, local absorption or the change of the refractive index leads to a contrast change in the image and enables the detection of the precipitates. The Si samples have to be polished on both sides to reduce the scattering of the light and therewith to enhance the resolution of IRM considerably. Hence, any commercial light microscope, which can be driven in transmission mode and which is equipped with a monochrome CCD-camera, can be used for determining precipitates in Si material. For taking IR images, the IR filter of the CCD camera has to be removed. Since the sample itself acts as a long-pass filter cutting all wavelengths below about 1100 nm, and the CCD-camera detects only light up to approximately 1100 nm, only a very narrow spectral range at about 1100 nm is used here. For our purpose we use a *Jenatech inspection* light microscope from Zeiss Jenoptik equipped with an *FK 6990-IQ* camera from Pieper. Furthermore, our microscope is equipped with a computer to control a step motor driven x-y table and to acquire and stitch the IRM images, which allows us to scan areas of sizes of approximately 5 cm × 5 cm in one run. For avoiding overload of the camera at the edges of the sample, where also unfiltered light reaches the camera, the whole sample is placed on a larger Si wafer polished on both sides, which acts as a filter also in these regions.

IRM Extended Focus Imaging

Extended Focus Imaging (EFI) is used to image objects sharply which are larger than the depth of focus of the microscope. For that purpose transmission IRM images in different focal planes within the sample are taken. Afterwards this image sequence is used to calculate an image in which the whole object is sharply pictured.

Electrical Measurements

It has to be distinguished between the measurements of $I-V$ characteristics of solar cells and the electrical measurements of precipitates in mc-Si material for solar cells.

Most of the $I-V$ characteristic measurements on solar cells in this work are performed with a *6632B* power supply from Agilent, which has an accuracy of $\Delta I = \pm 0.5$ mA for current measurements and a maximum power output of 100 W, i.e. 5 A at 20 V. This instrument is able to work as a current sink and in four-probe configuration and is equipped with a self-made trigger for the voltage [38], because it is also used to apply the pulsed voltage needed for DLIT and ILIT measurements.

The electrical measurements on precipitates are performed in an SEM *JEOL JSM 6400*, because the precipitates are of sizes in the range of some µm. Briefly, electrical two-point probe measurements at precipitates are performed by contacting the specimen by electrochemically etched platinum-iridium tips [66] by using two nanomanipulators (Kleindiek [67]) in the SEM. For a detailed description of that method the reader is referred to [68].

Chapter 3

Factors Influencing $I-V$ Characteristics of Solar Cells

In this chapter the characteristic features of the shape of real $I-V$ curves of Si solar cells are described. The physical mechanisms behind the individual effects are, if possible, approached on the basis of LIT, $I-V$ measurements, SEM, TEM, EBIC, and other experimental techniques.

To give a complete overview of the behavior of $I-V$ characteristics, a review of older and some recent work about effects in forward $I-V$ characteristics of mc-Si solar cells and their origins is given in sections 3.1.1 and 3.1.4. One major problem for the description of real forward $I-V$ characteristics is a strong discrepancy between the theoretical value of the ideality factor of $1 < n \leq 2$ [13] and its real value coming from the two-diode model fit procedure. Attempts to describe the physical origins of ideality factors larger than 2 as well as the effect of shunts on real forward $I-V$ curves are shown in section 3.1.4. The origins of different types of shunts are determined. Especially a detailed description of material-induced shunts caused by precipitates, their growth, and the shunt mechanism is given in section 3.1.3.

In section 3.2 new results on reverse $I-V$ characteristics of mc-Si solar cells are shown in detail. The term "pre-breakdown" relating to solar cells is concretized. By making use of the newly developed LIT methods to image the physical parameters of reverse-biased solar cells, a distinction of different pre-breakdown mechanisms is made in sections 3.2.2 to 3.2.5. The causes of the unexpectedly low (pre-) breakdown voltage in acidically textured solar cells are determined in detail, and first attempts to describe pre-breakdown phenomena theoretically are made.

3.1 Results: Forward $I-V$ Characteristics

In this section several physical mechanims which mainly influence the forward characteristic of solar cells and thereby affect the efficiency of solar cells negatively are given. Most of these effects have a strong influence on the parallel resistance and on the recombination current of solar cells. Leakage currents in forward $I-V$ characteristics of solar cells mainly come from edge recombination sites, shunts due to cracks or scratches and material-induced defects. It is important to note that contrary to previous assumptions for the two-diode model, LIT measurements show that all R_p and J_{02} currents, which will be discussed in this chapter, are local.

3.1.1 Linear Shunts

Linear shunts show a linear I–V curve behavior and affect the I–V curves of solar cells as shown in Figure 3.1. The curve displayed by the red circles in Figure 3.1 is the forward I–V curve of the linearly shunted solar cell shown later on in Figure 3.3. Linear shunts of any kind decrease the R_p of solar cells locally, i.e. at these sites an increased current in the solar cell flows. Therefore, linear shunts strongly reduce the efficiency of solar cells and must be avoided. Moreover, linear shunts are also leading to hot spots under reverse bias. Therefore, it is necessary to know the shunt mechanism in detail, down to their microscopic origins. In this section different origins of linear shunts in solar cells are summarized and explained.

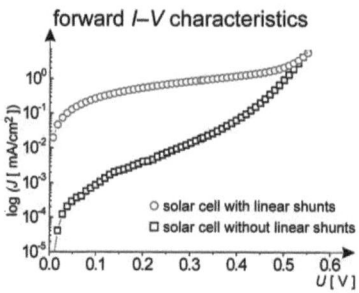

Figure 3.1: Comparision of an I–V characteristic measured at the solar cell which is shown in Figure 3.3 (red circles) and a solar cell without linear shunts (black squares).

Edge Shunts, Cracks, Aluminum Particles

Linear edge shunts are mainly caused by wrong laser or etching parameters during the isolation of the edges of a solar cell (see section 1.1 and Figure 1.1). If the laser cut or the edge etching is not deep enough, the n^+ layer has still an ohmic connection to the back side contact, which leads to a short circuit at the edges of the solar cell and produces linear shunts. The laser might also cut into the silver grid contact finger at the edge of the solar cell and burn the silver into the laser groove, which also may lead to a short circuit between emitter and back contact and therefore to linear shunts. These types of linear shunts are due to processing failures and can be easily avoided by choosing the right parameters for the respective process steps.

Cracks in solar cells may also lead to a severe decrease of R_p due to linear shunts. In Figure 3.2a) and b) DLIT images of a solar with linear shunts are shown. This shunt is caused by the crack shown in Figure 3.2c). The crack was already in the wafer before processing or it appeared during the processing due to handling. Whenever that crack happened, it occured before the screen printing step, i.e. step 5 described in section 1.1, took place. During the screen printing of the back side contact of the solar cell, the aluminum paste was pressed into the crack up to the front side of the solar cell. Since the crack crosses some silver grid lines at the front side of the solar cell, a direct short circuit between front and back metallization of the solar cell occurs. This can be seen in the light microscopy (LM) image in Figure 3.2c).

Note, that at a crack in an as-sawn wafer the crack edges will be n^+ doped by the POCl$_3$ diffusion step (step 2 in section 1.1) and may also make an ohmic contact to the p^+ layer. Therefore, also a very

3.1. RESULTS: FORWARD I–V CHARACTERISTICS

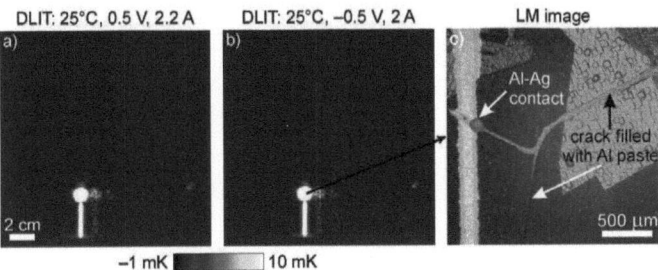

Figure 3.2: Linear shunt due to a crack in the Si wafer, which is filled with aluminum paste.

narrow crack, which maybe is so narrow that the silver paste can not pressed through the crack, may cause a linear shunt.

Another process induced-shunt was reported in [41]. There, aluminum particles, which get on the surface of the solar cell before firing due to contamination from other solar cells or due to spilled Al, lead to shunts. Al particles on the front side of a solar cell will locally form a p^+ layer by overcompensating the doping of the n^+ layer on the surface of the solar cell during firing. This p^+ layer is in direct contact to the p-doped base and forms an ohmic tunnel contact to the n^+ emitter leading to linear shunts in solar cells.

Material-Induced Shunts

In Figure 3.1 an I–V characteristic of an acidically texturized mc-Si solar cell with very low R_p (red circles) in comparison with an I–V characteristic of a good solar cell (black rectangles) of the same type is shown. The current of the shunted solar cell increases strongly at low voltages up to approximately 0.5 V according to the explanations given in subsection 1.4.1 (cf. Figure 1.9a). The overall parallel resistance of the shunted cell in Figure 3.1 is below 1 Ω, which is about three orders of magnitude lower than expected for mc-Si solar cells of this type. The low R_p comes from linear shunts, which are shown in the amplitude DLIT images in Figure 3.3. The comparison of the DLIT amplitude at the shunt position (see solid arrow in Figures 3.3a) and b) at 0.5 V forward and –0.5 V reverse bias, respectively, reveals that nearly all shunts are linear.

Please note that at the positions of the dashed arrows in Figure 3.3a) weak non-linear shunts are visible. As mentioned already above, the shunt current, i.e. the R_p current, is local. The specific sites are not clearly visible in the amplitude DLIT images in Figures 3.3a) and b) because of the poor spatial resolution. However, the $S^{0°}$ image in Figure 3.3c) taken at 0.5 V has a much better spatial resolution and shows that the shunts are point-like heat sources.

To reveal the origins of these linear shunts, the shunted area (marked by the white square in Figure 3.3c) was cut out of the solar cell with a dicing saw and then polished on both sides for performing infrared microscopy. In Figure 3.4a) the detailed $S^{0°}$ DLIT image of the shunted area, and in 3.4b) the corresponding IRM image of the section are shown. The dark features in the IRM image are Si_3N_4 inclusions and silicon carbide (SiC) precipitates in the mc-Si material [65, 68, 69]. In Figure 3.4c) an overlay of the $S^{0°}$ DLIT image and the IRM image is shown. The positions of the shunts coincide with positions of precipitates. At the shunt positions filament-like structures, which are arranged like a "fence", can be seen. An example is shown in Figure 3.4d), which displays a highly magnified IRM image of the area marked by the black solid rectangle in Figure 3.4c).

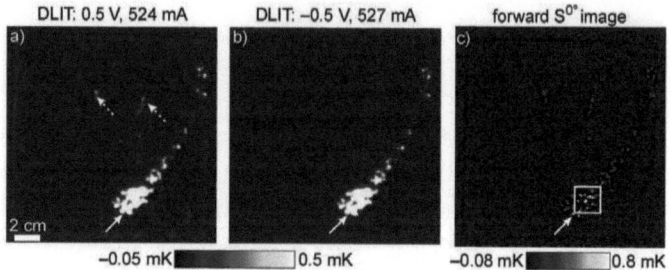

Figure 3.3: Solar cell with linear shunts. In a) and b) the amplitude DLIT images at ± 0.5 V, and in c) the $S^{0°}$ image of DLIT measurement at 0.5 V are shown.

3.1.2 SiC Filaments

The filament-shaped structures found at linear shunts are identified to be SiC filaments growing in the mc-Si material during the block-casting process. SiC filaments are frequently found in the upper part of mc-Si ingots. Since such SiC filaments may lead to severe ohmic shunts, it is necessary to find out more about the electrical and structural properties of these precipitates. In the last years the efforts on the investigations of precipitates in mc-Si of solar cells have increased. In this subsection an overview of former work and also new data are presentd. Besides the properties of SiC filaments, also their growth mechanism is of high interest for finding ways to avoid these dangerous precipitates and increase the yield of feedstock and solar cell production. The results of this work may be especially interesting for the new UMG material used for solar cells because of its higher amount of contaminations, which may lead to increased precipitation. Therefore, an idea about the growth of SiC filaments is presented. Furthermore IRM extended focus images are presented, which will help to reveal the shunt mechanism of SiC filaments in more detail.

Electrical Properties of SiC Filaments

Already in 2004 Al Rifai et al. [70] reported that the causes of shunts in mc-Si solar cells are SiC filaments. Later on the electrical properties of these precipitates were analyzed by Bauer et al. [68, 69]. The resistivity of the SiC filaments was found to be $\rho = 2 \times 10^{-3}$ Ωcm [68], hence they are highly conductive. Four-point probe measurements revealed that the SiC filaments are n-doped, and the charge carrier concentration was calculated to be $n_e \approx 8 \times 10^{18}$ cm^{-3} [68]. Since also a high amount of Si_3N_4 precipitates and Si_3N_4 inclusions is found in the vicinity of SiC precipitates, it was assumed that the doping of the SiC filaments is due to nitrogen, which is the main shallow donor for SiC [68]. In the course of this work we succeeded to measure the nitrogen contamination of the SiC precipitates by EDX qualitatively. For that purpose a TEM sample of a SiC particle was prepared by FIB. On this thin sample a line scan EDX analysis in a high resolution transmission electron microscope was performed. In Figure 3.5 the EDX line scan spectra along the red line in the inset of Figure 3.5 is shown. Although the EDX signal of nitrogen is very low, a significant increase of the nitrogen content from the Si to the SiC measureable. This is, to my knowledge, the first experimental proof that nitrogen is highly concentrated in SiC precipitates. From this it follows that the high conductivity of SiC precipitates in block-cast mc-Si materials is due to the high doping by nitrogen.

3.1. RESULTS: FORWARD I–V CHARACTERISTICS

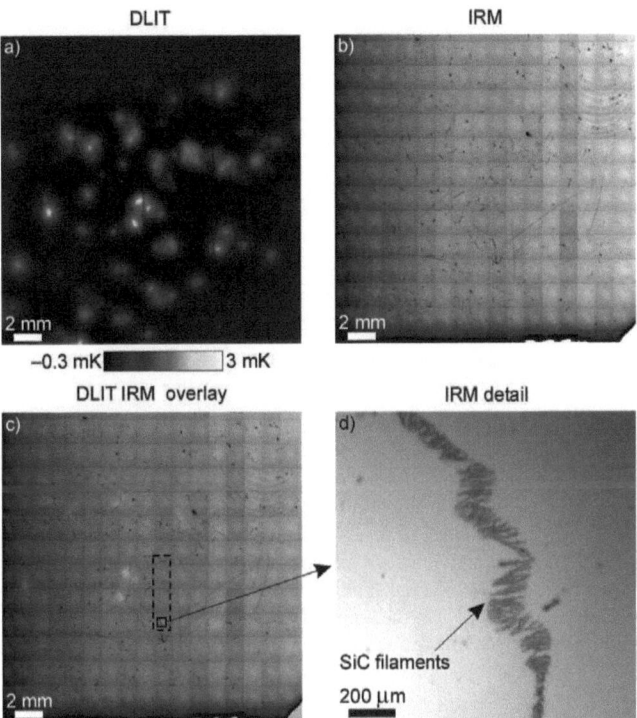

Figure 3.4: a) shows the $S^{0°}$ DLIT image of the area marked by the white quadrat in Figure 3.3c), b) is the corresponding IRM image, c) shows the overlay of a) and b). In d) the detail marked by the solid square in c) is shown. (The area marked by the dashed rectangle in c) will be shown in detail in Figure 3.4).

Structural Properties of SiC filaments

SiC has more than 200 polytypes and is a wide bandgap semiconductor, whose bandgap depends on the polytype. Former works revealed that the SiC filaments consists of cubic SiC (3C SiC) [41], with an $E_{\text{gap}} \approx 2.2$ eV [71].

There are different types of precipitates in mc-Si material: Si_3N_4 rods, Si_3N_4 filaments in grain boundaries (GB) and Si_3N_4 fibres. Furthermore, there are SiC crystallites which form SiC clusters, SiC filaments, which grow from these SiC crystallites but not in grain boundaries, and SiC filaments growing in grain boundaries of mc-Si material (which may also start growing from SiC clusters) [68, 69]. Only the latter two types of SiC precipitates, i.e. the SiC filaments, may cause linear shunts in solar cells. All Si_3N_4 precipitates are electrically insulating and therefore not responsible for linear shunts [68]. SiC crystallites are also highly conductive, but their size rarely exceeds the thickness of the solar cell, hence SiC clusters hardly produce short circuits in solar cells. Since SiC filaments in grain boundaries are the main cause of linear shunts [68], here only new results on the crystallographic structure of such SiC filaments are shown.

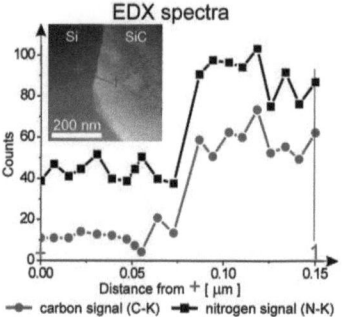

Figure 3.5: EDX line scan spectra. The spectra are taken from Si (cross in the inset) to SiC (the "1" in the inset).

SiC filaments in grain boundaries have an average diameter of about 1 to 3 μm and may grow some mm in length. The crystallographic structure of the SiC filaments in grain boundaries was investigated in much more detail in the present work than in [41], and was done by preparing cross- and longitudinal sections of the filaments by FIB. In this work we have observed that not only SiC filaments but also Si_3N_4 filaments may be present in grain boundaries side by side. In Figure 3.6a) an SEM image of a Si_3N_4 filament and SiC filaments sticking out of a grain boundary of mc-Si is shown. The FIB lamella (prepared at the marked filaments) of a longitudinal section parallel to the grain boundary is shown in Figure 3.6b). On the so-prepared samples it was possible to perform detailed TEM investigations and electron diffraction investigations to reveal the crystallographic structure of the filaments.

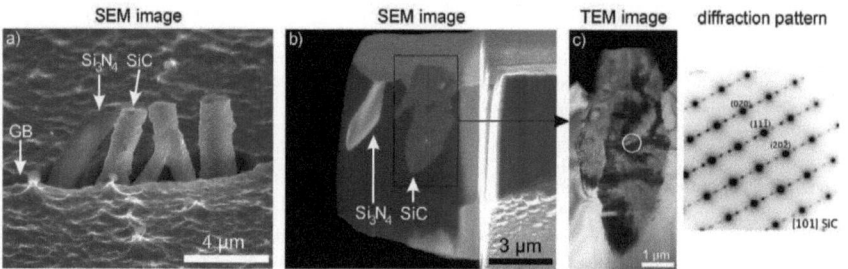

Figure 3.6: In a) filaments consisting of Si_3N_4 and SiC sticking out of a grain boundary (GB) are shown, b) shows an SEM image of the FIB lamella of the two left filaments. In c) a TEM image of the SiC filament and the electron diffraction pattern taken at the position marked by the circle is shown. The electron diffraction pattern reveals that the SiC filaments consist of 3C SiC, with a lattice constant of 0.435 nm (space group $F\bar{4}3m$). The spots between the main spots originate from twinning.

A TEM image of the SiC filament and the corresponding diffraction pattern taken at the area marked by the circle in Figure 3.6b) can be seen in Figure 3.6c). As can be seen in the TEM image, the SiC filament is polycrystalline. Its structure is disturbed by crystallographic defects like stacking faults and twin boundaries.

3.1. RESULTS: FORWARD I–V CHARACTERISTICS

Furthermore, the investigation by electron diffraction (see Figure 3.6c)) shows that the SiC filaments consist of 3C SiC. All these results are published in [72], where also more details on SiC crystallites and Si_3N_4 fibres can be found.

Shunt Mechanism or the Origin of Very Low R_p of Solar Cells

As mentioned above, SiC filaments grow in the direction of the solidification of the mc-Si block (cf. Figure 3.8 below). Therefore, they are orientated perpendicular to the solar cell surface. Since SiC filaments may be some mm long, they exceed the thickness of a solar cell several times [69], and will stick out on both sides of the Si wafer used for the solar cell production. Because of their high conductivity, SiC filaments may cause a short circuit between the emitter (front contact) and the back contact of the solar leading to severe linear shunts. Note that the filaments are n-conducting and show only a weak band offset to the conductive band of silicon [68]. Therefore, they are in direct contact to the emitter and "extend" the emitter into the depth, which can nicely be observed by back side EBIC imaging [61]. Only if a filament is in ohmic contact to the metallization, it produces a shunt.

The length of a SiC filament in a solar cell equals the thickness of the Si wafer and is approximately 200 μm. Regarding this length, a filament diameter of about 1 μm, a resistivity of $\rho = 2 \times 10^{-3}$ Ωcm, and a cylindrical shape of the SiC filaments, one SiC filament has a resistance of about 710 Ω [68]. That means if there are only ten SiC filaments making parallel short circuits in a solar cell, the R_p of this solar cell is well below 100 Ω, which already makes the solar cell degraded.

However, during the investigations the observation was made that some grain boundaries, which contain SiC filaments, cause only weak or even no shunts at all. In Figure 3.7a) an example of a very weak shunt and a strong shunt caused by SiC filaments from the area marked by the dotted rectangle in Figure 3.4c) is shown.

In Figure 3.7d) another overlay image of an $S^{0°}$ DLIT image and an IRM image of a different solar cell is shown just to demonstrate that this phenomenon appears frequently and is worth to be investigated more carefully with regard to understanding the shunt mechanism in detail and maybe find ways to avoid linear shunts due to SiC precipitates.

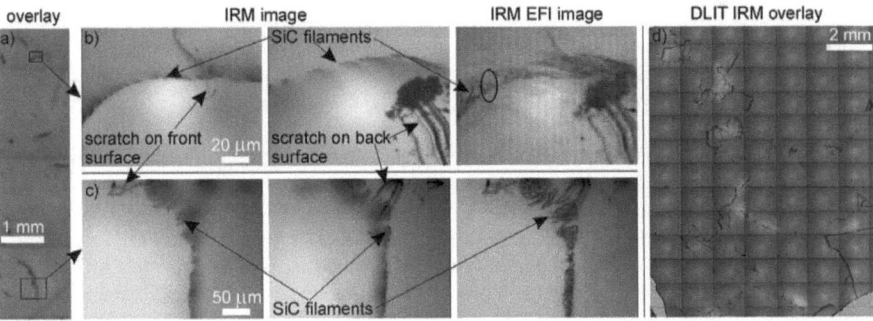

Figure 3.7: In a) the section marked in Figure 3.4 is shown, the dark lines are SiC filaments in grain boundaries, b) and c) show IRM and IRM extended focus images of SiC filaments. A detailed explanation can be found in the text.

The dark lines in Figure 3.7a) and d) are large angle grain boundaries, which are partly contaminated with SiC filaments. The bright spots are DLIT signals of linear shunts, but not at every contaminated grain boundary a DLIT signal due to a shunt is measureable, or at least a significantly lower DLIT signal is measured. The SiC filaments at the grain boundaries in the lower right corner and on the lower left side in Figure 3.7d) do not show a DLIT signal at all.

The question is, why do some grain boundaries containing SiC filaments cause shunts and others do not? The mechanism behind that behavior is not fully understood yet. However, first investigations on shunting filaments and on non-shunting filaments by IRM extended focus imaging show that there are some geometrical differences, which can cause different shunt behavior [73]. This is demonstrated exemplarily in Figure 3.7b) and c). The left and the middle pictures in b) and c) show IRM images of the front and the back surface of the sections marked in a), respectively.

For better orientation scratches were made on each surface of the wafer near the precipitates by a diamond scriber. These scratches are marked in Figure 3.7. The first images in b) and c) show single IRM images for two different focus positions (top and bottom), where only parts of the filaments are sharply imaged due to limited depth of focus. The right images in b) and c) show the extend focus IRM images, which reveal the complete shape of each single SiC filament through the Si wafer.

The SiC filaments in Figure 3.7b) show only a very weak shunt signal. As can be seen in the IRM images in b) not all SiC filaments which start at the surface in the left image can be found in the middle IRM image, i.e. on the opposite wafer surface. From this it follows that some of the filaments are interrupted within the wafer, which exemplarily can be seen at the ellipse in the EFI image of Figure 3.7b), or branches out. Hence, some of the SiC filaments do not reach the other surface. Therefore, they can not lead to a short circuit between front and back contact. Obviously this effect leads to an increased resistance of the whole contaminated grain boundary compared to other contaminated grain boundaries. Therefore, the shunt current at this specific site in the solar cell is lower in comparision to other shunt sites and the DLIT signal here is smaller.

In Figure 3.7c) nearly all filaments penetrate the Si wafer. This is proven by the left and middle image, where nearly all SiC filaments, which enter the surface on the one surface, can be found also in the other surface. The EFI image shows that the filaments go straight through the wafer and only a few are interrupted.

It has to be mentioned here that the density of SiC filaments in the grain boundaries, i.e. the amount of SiC filaments per unit length, in the left images of Figure 3.7b) and c) is nearly the same (please note the different scales). But, at least in this example, there is a significant difference in the amount of filaments which reach the other surface of the wafer (see middle image of b) and c)). Especially in the left part of the middle image of b) the density of SiC filaments decreased strongly compared to the same region in the left image of b)). However, it has been found that there are also examples where nearly all filaments are grown through and nevertheless do not produce a shunt.

At the moment there is no explanation why the SiC filaments grow differently or what causes an interruption of their growth. It is also not known yet under which conditions the filaments have ohmic contact to the back side metallization. Some hypotheses about the growth of SiC filaments in mc-Si have been made by S. Möller et al. [74], and H.J. Möller [75] but there has been no complete model yet. Though, such a model is required to explain the differences of the growth of SiC filaments in mc-Si material. In the next subsection a deeper insight into the growth of SiC filaments in grain boundaries of mc-Si is given.

3.1.3 Growth of SiC Filaments in Block-Cast mc-Si

Regarding the structural properties of the SiC filaments and the results of IRM, SEM and TEM investigations given in the previous subsections and in the publications cited there, assumptions of the

3.1. RESULTS: FORWARD I–V CHARACTERISTICS

growth of SiC filaments will be made in the following section. For explaining the growth of the SiC filaments it is important to consider the block-casting process of the mc-Si ingots (which was already mentioned in section 1.1), the properties of carbon (C) in liquid and solid silicon, as well as the appearance and properties of the SiC filaments.

Properties of the Block-Casting Process

In Figure 3.8a) a typical solidification or temperature–time (T–t) diagram of the solidification of an mc-Si block is shown schematically. After the Si feedstock is melted completely in the crucible ($T_{\text{melt Si}} = 1683$ K) the temperature at the bottom of the crucible is decreased and the Si starts to crystallize from the bottom to the top of the crucible. After the whole ingot is crystallized, the temperature is decreased faster, resulting in a steeper temperature ramp as shown in Figure 3.8a). In Figure 3.8b)

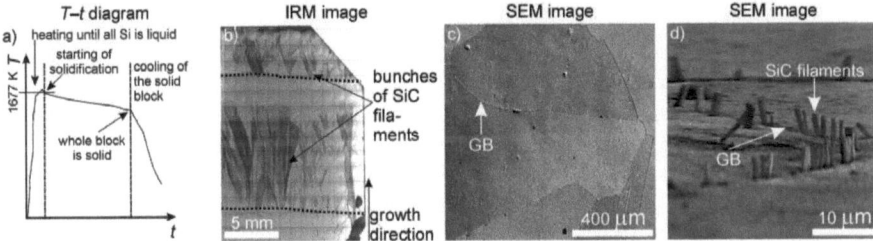

Figure 3.8: In a) a typical solidification diagram (T-t) of a point at the bottom of an mc-Si ingot is shown schematically. The vertical cut of the top part of an mc-Si ingot in b) shows that at the top of mc-Si ingots often SiC filaments can be found. They grow several mm in length in direction of the growth of the ingot (see arrow). In c) the etched surface of an mc-Si wafer containing grain boundaries is shown, d) displays a magnified image of a GB of c) and reveals SiC filaments in the GB.

a vertical cut of the uppermost part of an mc-Si ingot containing SiC filaments is shown. The SiC filaments grow towards the solidification direction of the ingot and bunches of SiC filaments start at different heights, which is illustrated and marked by the dotted lines in Figure 3.8b). That cascade-like behavior of the precipitation of SiC filaments leads to the conclusion that the carbon concentration in the Si starts to drop during the growth of the filaments until the carbon concentration falls below the limit where no further growth of SiC is possible. After the formation of SiC filaments has stopped, the carbon concentration rises again to the concentration level needed for further SiC precipitation.

The carbon contamination of the mc-Si comes mainly from the crucible mountings which are made from graphite. Carbon sources are carbon monoxide (CO) and carbon dioxide (CO_2). CO and CO_2 are formed at the hot graphite parts in the block-casting machine if there is residual water in the system. The water vapourizes during the heating of the crucible and reacts with the hot graphite to CO and CO_2. If CO and CO_2 reach the liquid Si they will be cracked and carbon is dissolved in the liquid Si. Moreover, it is known that CO is formed at the contact of the quartz crucible with the hot graphite [76]. Other possible carbon sources are carbon contamination of the feedstock and of the Si_3N_4 covering of the crucible, which slowly dissolves during crystallization.

Properties of Carbon and SiC Filaments in Silicon

During solidification the amount of liquid Si in the crucible decreases as the solid fraction increases. Depending on their segregation coefficient, foreign atoms are concentrated in the liquid Si or preferentially build into the solid fraction. Carbon (C) has a segregation coefficient of 6×10^{-2} in silicon [77] and will be concentrated in the liquid phase. The concentration of carbon in solid silicon is the concentration of carbon at the solidification temperature. The values found in literature are very different. The concentration of C in solid silicon at its melting point was measured by Nozaki et al. to be 3.5×10^{17} atoms/cm^3 [78]. In [80] the concentration of carbon in solid Si at temperatures from 1400 °C to 1200 °C was given to be approximately 10^{-3} at.%, i.e. 5×10^{19} cm^{-3}. Søiland et al. measured a carbon concentration in the solid Si of an mc-Si block for solar cells of 14 ppma, i.e. 7×10^{17} cm^{-3}. Furthermore, Kalejs et al. [81] reported a carbon concentration of 8.5×10^{17} cm^{-3} in solid silicon. Therefore, the value in [80] was considered to be wrong and a value in the order of 10^{17} cm^{-3} should be used. By measuring the carbon content in the liquid and solid phase of industrial mc-Si ingots, Søiland et al. found that the carbon content in the liquid fraction of the ingot is well below its solubility limit at the beginning of the solidification. The liquid fraction becomes smaller during solidification and therefore the carbon concentration increases, reaches a maximum and then drops, since SiC is forming in the melt [79].

At the beginning of the casting process the carbon content in the solid fraction is also below the solubility limit. During the block-casting process the carbon concentration in the solid Si increases together with the carbon in the liquid fraction. After a certain time the carbon concentration reaches a maximum, i.e. the carbon is supersaturated in the liquid fraction and saturated in the solid fraction. If the supersaturation is reached, SiC precipitates, which leads to the drop in the carbon concentration in the liquid as well as in the solid phase [79]. From these circumstances one can conclude that SiC is formed at least in the liquid but maybe also in the solid phase of silicon. Indeed SiC clusters can be found frequently in mc-Si ingots, whose perfect crystallographic structure leads to the assumption that they grow in the liquid Si [69, 72].

In Figure 3.8c) and d) the surface of an mc-Si wafer containing SiC filaments is shown. The wafer was polished, and afterwards about 5 μm of the surface was etched away by a mixture of hydrofluoric acid (HF) and nitric acid (HNO$_3$). Since SiC is inert against this etching solution, the SiC filaments remain and stick out of the grain boundary. The grain boundaries can be nicely seen in Figure 3.8c), because they are preferentially etched by the HF+HNO$_3$ mixture. In Figure 3.8d) a magnified SEM image of a GB in c) is shown. All SiC filaments are arranged in the grain boundary.

Most of the SiC filaments are found in grain boundaries of the mc-Si material, but often not the whole grain boundary is contaminated with SiC filaments. Hence, one can conclude that the grain boundaries are not caused by the SiC filaments, they must have been formed already before the SiC filaments have started to grow. Due to energetic reasons grain boundaries are favored sites for the precipitation of SiC in a Si matrix. As demonstrated in Figure 3.6 the filaments are polycrystalline and their crystallographic structure is heavily disturbed. The reason for the imperfect crystallographic structure of the filaments is assumed to be the growth of the filaments inside the already formed grain boundary. This assumption is supported by the fact that the filaments do not show a defined growth direction along a crystallographic direction, but follow the direction of the grain boundary. Furthermore, the interface between the SiC filaments and the surrounding Si matrix is wavy and rough [72].

Growth Models

In principle there are two possible modes of growth for SiC filaments in grain boundaries of mc-Si material. The first approach by H.J. Möller et al. is to assume that the SiC filaments are formed in the liquid phase, i.e. above the liquid–solid interface, during the block-casting process [82]. The model by

3.1. RESULTS: FORWARD I–V CHARACTERISTICS

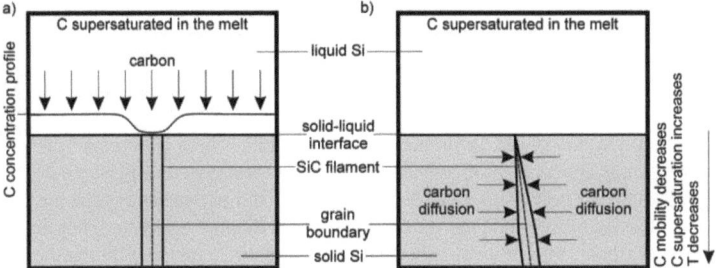

Figure 3.9: In a) the model of the growth of the SiC at the phase interface and in b) the solid diffusion model are shown.

H.J. Möller et al. is shown schematically in Figure 3.9a). Briefly, the carbon concentration c_C in front of the liquid–solid interface has a certain profile because carbon atoms are incorporated in the solid Si. If SiC precipitates e.g. in a grain boundary the carbon concentration is reduced there. Assuming that there is enough carbon in the liquid Si the growth of the SiC crystallites will continue and a SiC filament will be formed as it is illustrated in Figure 3.9a). Note that the model of Möller et al. [82] is based on experimental data of SiC filaments which do not grow in GB but within a grain of mc-Si.

The second approach to explain the formation of SiC filaments is the diffusion of carbon atoms in solid silicon. A detailed description of that model is given in the next section.

Solid State Diffusion Growth

In Figure 3.9b) the solid state diffusion model is shown schematically. Regarding the facts that SiC filaments are frequently found in grain boundaries and that their crystallographic structure is imperfect, it is assumed in this work that the growth of the SiC filaments takes place in the solid silicon.

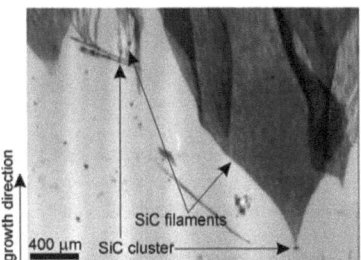

Figure 3.10: IRM image of SiC filaments growing from different nucleation sites. The filaments on the right start to grow from a single SiC cluster which can be seen in the bottom right corner. At the top left a SiC cluster which itself is grown on a Si_3N_4 rod acts as nucleation site for a SiC filament.

Therefore, solid state diffusion of carbon towards grain boundaries of mc-Si has to be assumed for the formation of the filaments. In Figure 3.10 an IRM image of a bunch of SiC filaments is shown. A SiC filament starts to grow at a nucleation point, which here is a small SiC cluster. The filament branches out and a "curtain" of SiC filaments emerges from that first SiC filament. The filaments in the "curtain" grow in grain boundaries and their growth starts closely beneath the solid–liquid interface until the filament reaches its typical diameter, as it is shown in Figure 3.9b).

The question is whether solid state diffusion of carbon atoms in silicon is fast enough to transport the required amount of C atoms to the nucleation site (e.g. a grain boundary containing SiC) to form SiC filaments by regarding the parameters of the block-casting process?

To answer that question firstly the amount of carbon atoms which is necessary to form SiC filaments as they typically appear in grain boundaries of mc-Si is estimated. Secondly a detailed description of the block-casting process and the determination of the parameters which are important for estimating the diffusion of carbon in the solid fraction of the Si ingot are given. Then an estimation of the change of the carbon concentration c_C in the vicinity of grain boundaries during the block-casting process is made by solving the diffusion equation to check whether solid state diffusion is a probable mechanism to explain the growth of SiC filaments.

Estimation of the Amount of Needed Carbon Atoms

For the estimation of the amount of carbon atoms needed for SiC filaments experimental data are used. In Figure 3.11 a model grain boundary occupied with SiC filaments is shown schematically.

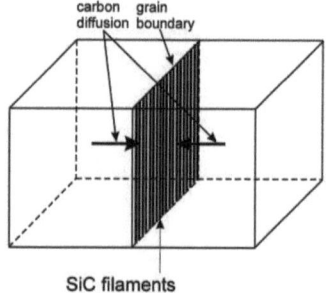

Figure 3.11: Schematical view of a grain boundary with SiC filaments. Please note that the sketch is not to scale.

It is assumed that SiC filaments in grain boundaries have an average diameter of 1 μm, and that their typical average density in a grain boundary is approximately 900 filaments per cm (average distance of about 11 μm). Assuming further a cylindrical shape of the filaments with a volume of $V_{filament} = \pi(d/2)^2 l$, where l is the length of the filaments (here 1 cm), the volume of all SiC filaments in the unit area of 1 cm² is $V \approx 7 \times 10^{15}$ nm³. The lattice constant of the unit cell of 3C SiC is $a = 0.43596$ nm. By dividing V by the volume of the SiC unit cell it follows that about 8.5×10^{16} SiC unit cells are needed for the volume of SiC filaments typically found in grain boundaries in 1 cm². Four atoms per SiC unit cell are carbon atoms, which means that

$$N_{carbon} = 3.4 \times 10^{17} \text{cm}^{-2} \tag{3.1}$$

carbon atoms are needed to form SiC filaments typically found in grain boundaries of mc-Si.

3.1. RESULTS: FORWARD I–V CHARACTERISTICS

Parameters of the Block-Casting Process

The solidifcation front of the block is assumed to be horizontal and a cooling rate κ of the ingot is defined as

$$\kappa(t) = \frac{dT(t)}{dt}, \quad (3.2)$$

which gives the change in temperature T at a certain position in the ingot with time t. Note that κ is negative. For simplification the cooling rate of the block-casting process for mc-Si is assumed to be constant in certain time intervals, i.e. the temperature–time profile of a certain position in the ingot is linear. Then equation (3.2) simplifies to

$$\kappa = \frac{\Delta T}{\Delta t}, \quad (3.3)$$

with ΔT the change in temperature during a certain time intervall Δt. This simplification is in fact not a too rough estimation as can be seen in Figure 3.8a). However, it is appropriate to split the temperature time profile into several linear regions with different cooling rates κ_i as it is shown below (cf. Figure 3.12).

In Figure 3.12 a typical solidification diagram is shown. Two ranges of cooling with constant cooling rates κ_1 and κ_2, respectively, are assumed. That is a slow range, which holds until all Si is solidified, and a fast range, which holds when all Si is solidified, as can be seen in Figure 3.12. Typically SiC filaments are found at about 85 % of the overall height of the ingot up to the top of the

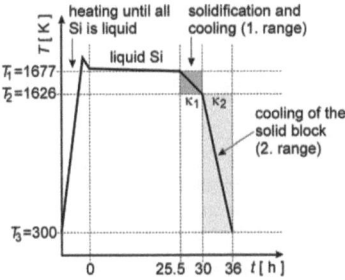

Figure 3.12: Schematical solidification profile of an mc-Si ingot at the "85 % position".

ingot. Therefore, their growth starts not till then if 85 % of the ingot is already solidified. That means the time when the SiC growth starts is 85 % of the overall cooling time. In the example given here the time until all Si is solid is 30 h which is a realistic assumption for small mc-Si blocks. The time for cooling down the block after full solidification is approximately 6 h as can be seen in Figure 3.12.

For a typical solidification process of an mc-Si block an average solidification velocity of $v_s = 1.5$ cm/h as well as a typical temperature gradient in the solid Si of $\nabla T = -7.5$ K/cm can be assumed. From this it follows that the cooling rate κ_1 of the solid fraction of the ingot is

$$\kappa_1 = v_s \nabla T = -11.25 \text{ K/h} = -0.0031 \text{ K/s} \quad (3.4)$$

until all Si is solidified. Since 85 % of 30 h are 25.5 h, i.e. 4.5 h are left for cooling the "85 % position" at κ_1, then the cooling rate changes to κ_2. The solid state diffusion starts at $T_0^{(1)} = 1677$ K, because then the Si is solid [80]. At the "85 % position" the Si is liquid until the solidification front reaches that position after 25.5 h as can be seen in Figure 3.12. For the typical cooling rate of $\kappa_1 = -0.0031$ K/s the "85 % position" cools down to $T_0^{(2)} = 1626$ K in 4,5 h. Therefore $t_1 = 16200$ s. Afterwards the solid ingot is cooled down to $T_3 = 300$ K in 6 h, from this it follows that

$$\kappa_2 = \frac{\Delta T}{\Delta t} = \frac{(300 - 1626) \text{ K}}{6 \text{ h}} = -0.0614 \text{ K/s}. \quad (3.5)$$

To prove whether enough carbon atoms are able to diffuse by solid state diffusion to the grain boundary to form SiC filaments, the change of the carbon concentration in the vicinity of the grain boundary is calculated. For that purpose it is assumed that the carbon concentration c_C towards the grain bound-

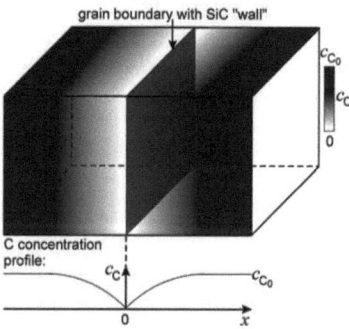

Figure 3.13: Schematical distribution of C concentration in the vicinity of a grain boundary containing a 2-dimensional SiC "wall".

ary decreases, as it is shown in Figure 3.13, because carbon atoms reaching the grain boundary are used for SiC formation. Hence, it is assumed that at the solid-liquid interface carbon is incorporated at the solubility limit. As soon as the temperature decreases, the solubility of carbon in the solid Si decreases, which leads to an increasing supersaturation of carbon in the solid silicon. The amount of carbon is then above the solubility limit in solid Si, which is the thermodynamic driving force for precipitation. At an adequate horizontal distance from the grain boundary the carbon concentration has the constant value c_{C_0}, which equals the solubility concentration of C in solid Si at the melting point. As indicated in Figures 3.11 and 3.13 GBs are favored sites for the nucleation of SiC precipitates. If SiC nuclei exist they will start to grow and consume C atoms out of their vicinity (this is indicated by the decreasing concentration towards the GB in Figure 3.13) if the carbon is supersaturated there.

Values of the diffusivity $D(T)$ for carbon in solid silicon can be found for instance in [81, 83, 84, 85]. All these publications cover more or less the same temperature range of about 1300 K to 1673 K. Here, the diffusivity found by Newman et al. is taken [83]:

$$D(T) \approx 0.33 \cdot e^{\frac{-2.92 \text{eV}}{kT}} \frac{\text{cm}^2}{\text{s}}. \quad (3.6)$$

For the estimation of the change of the carbon concentration c_C in the vicinity of a GB in mc-Si the block-casting process has to be considered. The change in temperature of a certain ingot position during the block-cast process with time t is described by κ and the temperature T_0 at which the cooling at that position starts:

$$T(t) = T_0 + \kappa t, \quad (3.7)$$

3.1. RESULTS: FORWARD I–V CHARACTERISTICS

by substituting T in equation (3.6) by equation (3.7) the diffusivity becomes a function of t:

$$D(t) \approx 0.33 \cdot \exp\left(\frac{-2.92\text{eV}}{k(T_0 + \kappa t)}\right) \frac{\text{cm}^2}{\text{s}}. \tag{3.8}$$

Since two stages of cooling with different cooling rates and durations are assumed, the parameters for each cooling range have to be considered. For the first range $\kappa_1 = -0.0031$ K/s, $T_0^{(1)} = 1677$ K, and $t_1 = 16200$ s and for the second range $\kappa_2 = -0.061$ K/s, and $T_0^{(2)} = 1626$ K holds. Since for $D(T)$ there are only valid values down to 1300 K, t_2 calculates to $t_2 = \frac{\Delta T}{\kappa_2} = \frac{(1300-1626)\text{ K}}{-0.061\text{ K/s}} \approx 5300$ s.[1] Hence, the overall cooling time t_{cooling} of the "85 % position" is 21500 s.

Estimation of the Change of c_C

To estimate how many carbon atoms are diffused towards the grain boundary to form SiC filaments one has to calculate the change of the concentration of carbon atoms in the vicinity of a grain boundary by solving the diffusion equation for that special problem. For that purpose the model grain boundary as shown above in Figure 3.13 was used. Since it was assumed that the carbon atoms diffuse only from the left and the right to the grain boundary the diffusion problem simplifies to a one-dimensional one. The one-dimensional diffusion equation is

$$\frac{\partial c_C(x,t)}{\partial t} = D(t)\frac{\partial^2 c_C(x,t)}{\partial x^2}. \tag{3.9}$$

In the model it is assumed that at the beginning there is a homogenous carbon concentration of the solubility limit in the solid silicon. Futher it is assumed that after solidification the solubility of the carbon in the cooling solid silicon reduced very fast due to its temperature dependence. To simplify the calculation it is assumed that c_C at the grain boundary approaches zero and c_C far away from the grain boundary is $c_{C_0} = $ constant. Then the boundary conditions for solving the diffusion equation (3.9) read as follows:
$\forall t \; c(0,t) = 0$ and $c(x_\infty,t) = c_{C_0}$, and the starting condition is: $c(x,0) = c_{C_0}$.
Equation (3.9) was solved numerically with $c_{C_0} = 7 \times 10^{17}$ cm^{-3}, which was measured in mc-Si for solar cells by Søiland et al. [79], and $D(t)$ from equation (3.8) under consideration of both cooling ranges and the cooling time t_{cooling}. Note that this assumption only gives an upper limit of the change of the carbon concentration in the vicinity of the grain boundary. Figure 3.14a) shows the $T(t)$ curve and the $D(t)$ curve of the block-casting process at the "85 % position". In the first cooling range ($\kappa_1 = -0.0031$ K/s) the diffusivity decreases only moderately whereas in the second cooling range ($\kappa_2 = -0.061$ K/s) the diffusivity decreases very fast, which means that the carbon atoms are "freezed" in the solid Si. In Figure 3.14b) the carbon concentration profile is shown. To estimate the amount of carbon atoms which are incorporated in the grain boundary, the integral over the hatched area in Figure 3.14 is calculated. The result is an areal density of carbon atoms of 1.2×10^{16} carbon atoms per cm^2. Since the carbon diffuses from both sides of the grain boundary the overall area density of carbon atoms is about 2.4×10^{16} cm^{-2}. From this it follows that 2.4×10^{16} carbon atoms diffuse in the unit volume of 1 cm^3 to the grain boundary during the block-casting process at the "85 % position" which is more than a factor of 10 less than the required number of $N_{\text{carbon}} = 3.4 \times 10^{17}$ cm^{-2}.

---
[1] Since the cooling rate in the 2nd region is quite high, the diffusion of C atoms is very slow and the mistake made by skipping the diffusion at temperatures lower than 1300 K is negligible.

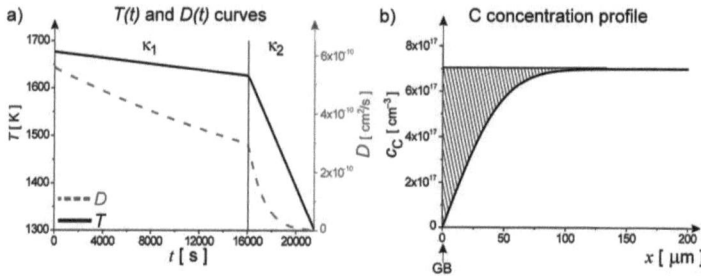

Figure 3.14: In a) the temperature and diffusivity curves of the block-casting process at the "85 % position" are shown. In b) the calculated carbon concentration profile is shown. The hatched area marks the amount of carbon atoms which are incorporated in the grain boundary.

Discussion

A strong argument for growth of SiC filaments in grain boundaries by solid state diffusion is the observed poor crystallographic quality [72] of such SiC filaments. The calculation above revealed that indeed not enough carbon atoms can be transported to the grain boundary to form the observed SiC filaments by classical solid-state diffusion. However, in spite of this result solid-state diffusion might be a feasible process, if one considers mechanisms which lead to enhanced carbon diffusion. In [81] it was reported that carbon diffusion in solid Si is enhanced due to silicon interstitials. The diffusivity was assumed to be a factor 10 higher than in Si samples without interstitials. If during the SiC precipitation silicon interstitials should be emitted or are existent in the silicon for other reasons in the vicinity of the SiC precipitiates they may enhance the carbon diffusivity. Another mechanism which may be feasible for the growth of SiC filaments in GB is the diffusion of carbon atoms in the GB. The diffusion in GB is much faster than diffusion in the bulk and may be a promising candidate to explain the growth in GB of mc-Si, if the carbon atoms diffuses from the melt in the GB to the SiC filaments. In order to explain the observed poor crystallographic quality of the SiC filaments in GB also a combination of the model of Möller et al. [75, 82] and the solid-state diffusion model is possible. Then most of the SiC grows at the liquid-solid interface and further carbon may diffuse towards the grain boundary in the solid silicon due to solid-state diffusion, leading to the bad crystallographic quality of the filaments.

The estimations made above are only the starting point to understand the growth of SiC filaments in mc-Si completely. However, a detailed description of the growth mechanism of SiC is needed to think about measures to avoid such precipitations, in particular if possibly UMG silicon is used for solar cell production in the future.

3.1.4 Non-Linear Shunts

In Figure 3.15 an example of non-linear shunts in a solar cell is shown. As can be seen in Figure 3.15a) and b) the non-linear shunts are only visible if the solar cell is forward biased. In this section a brief review on the origins of non-linear shunts and their role for n-factors of silicon solar cells larger than 2 is given. Figure 3.15 is the same Figure as Figure 2.4 in chapter 2, but here the scaling is different to highlight the non-linear shunts. The non-linear shunts obviously show frequently n-factors larger than 2 as can be seen in the n-factor map in Figure 3.15c). Note that the weak extended signal in the

3.1. RESULTS: FORWARD I–V CHARACTERISTICS

area of Figure 3.15a) is due maxima of the diffusion current at low lifetime regions of the solar cells. According to theory the n-factor in these regions is close to 1.

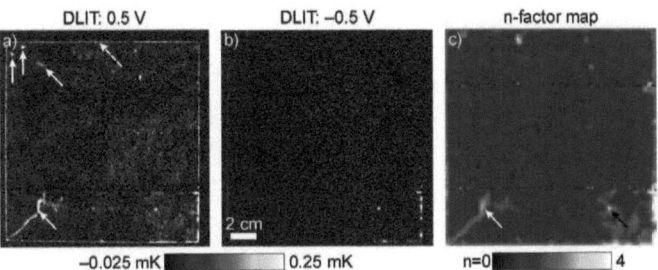

Figure 3.15: In a) and b) DLIT images at different voltages are shown. The non-linear shunts are marked by arrows. In c) the n-factor map of the cell is shown.

Recombination-Induced Shunts at Edges and Scratches

Non-linear shunts show increased recombination currents. It should be mentioned explicitly that these increased currents are local currents. In particular the edges of solar cells show increased recombination currents in forward bias direction as can be seen in Figure 3.15 a) (marked by the dashed arrow). In [26] it was shown that the recombination fraction of the forward current density increases the smaller the solar cell is. It was concluded that the recombination currents mainly flow at the edges of a solar cell. This is demonstrated in Figure 3.15a): at the edges of the solar cell high DLIT signals appear and at these sites an ideality factor of about 2 was measured. The non-linear shunt at the bottom left in Figure 3.15 is a scratch. This scratch shows an ideality factor larger than 2. In [86] it was shown that such scratches show highly increased recombination currents. Furthermore, it was shown in [86] that the n-factor of such scratch-induced non-linear shunts is larger than 2 and that n is increasing with the depth of the scratch.

The origin of the enhanced recombination current are extended defects. At scratches the defects are introduced due to plastic deformation of the silicon material, whereas at the edges of solar cells the defects are introduced due to the opening of the p-n junction by the laser cut or by chemical plasma etching. A short review of previous models explaining the recombination current was already given in section 1.4.2. Here, an overview of a new explanation for the enhanced recombination current, called coupled defect level (CDL) recombination [87], will be given. This model is partly based on the model of Schenk and Krumbein who considered recombination via two coupled defects [25]. However, Schenk and Krumbein had to assume an unrealistically high coupling of the defect levels.

If it is assumed that the recombination current flows locally at the extended defects, a high local density of defects is realistic. Laser cuts, scratches, or interfaces of precipitates may easily have defect densities up to 10^{15} cm^{-2} leading to interaction of the defects, because the distance between the defects is then well below 1 nm. The n-factor was simulated for a 10 µm wide highly defective region in a perfect p-n junction. The simulation[2] was performed with different coupling rates r_{12} between

---

[2]The simulation was done by P.P. Altermatt with the device simulator *Sentaurus* [88].

the two defects, for different defect energy levels E_1 and E_2, and taking into account the lifetime of a charge carrier at the defects $\tau_1 = \tau_2 = \tau$.

In Figure 3.16a) the measurements of n-factors of nearly perfect passivated emitter, rear locally diffused (PERL) solar cells which were treated with scratches of different strength or which were laser cut or indented are shown. In Figure 3.16b) and c) simulated n-factor curves are shown. In b) the coupling rates were kept constant, whereas in c) the n-factor was simulated for different coupling rates for two given constant lifetimes. The energy levels E_1 and E_2 are referred to midgap. For the

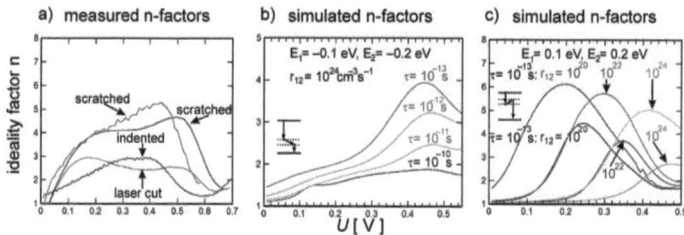

Figure 3.16: In a) measured n-factors of PERL solar cells are shown. In b) and c) simulated n-factors assuming a scratch in a PERL solar cell are presented. The Figure is taken from [87].

simulation of the case that $E_1 > E_2$ and constant coupling rate of $r_{12} = 10^{24}$ cm^{-3}s^{-1}, it turned out that the ideality factor has a peak at voltages between 0.3 V and 0.5 V, as can be seen in Figure 3.16b). For small lifetimes of $\tau = 10^{-13}$ s the ideality factor reaches values of 4 at 0.45 V, whereas the ideality factor for a lifetime of $\tau = 10^{-10}$ s in the voltage range between 0.3 V and 0.5 V is about 1.5. If $E_1 < E_2$ and the lifetime is kept constant (as shown in Figure 3.16c) the ideality factor depends strongly on the coupling rate. For small coupling rates of $r_{12} = 10^{20}$ cm^{-3}s^{-1} the ideality factor peaks at low voltages of about 0.15 V, whereas n has a peak at 0.3 V for higher coupling rates of $r_{12} = 10^{22}$ cm^{-3}s^{-1}, and so on.

The correlation between the parameters r_{12}, τ, and $E_{1/2}$ is complex, but the important result of the simulation is that only some parameters have a strong influence on the ideality factor. If the second defect level is at a higher energy than the first defect level ($E_1 < E_2$) the peak of n depends strongly on the coupling rate r_{12}. Furthermore, n increases with smaller τ, but the position where n has a peak are insensitive to τ and $E_{1/2}$. Experimental n-curves depend on a large variation of defect pairs and may be reconstructed by adding the simulated curves over a spatial and energetic range of defect pairs [87].

Due to the variation of the distribution of defect pairs the variation of the ideality factor varies for every experiment. Nevertheless, some conclusions can be drawn from the simulation. For example, the measured n-factors show a peak at about 0.4 V to 0.5 V, by comparing this with the simulation (cf. Figure 3.16b) this leads to the assumption that the energy level of the second defect is lower than that of the first defect. In conclusion the recombination current must be simulated beyond SRH theory and the CDL model seems to be able to describe enhanced recombination in solar cells sufficiently.

Schottky-Type Shunts

Schottky-types shunts have been described for instance in [41]. Schottky-type shunts occur if the front contact comes in direct contact to the p-doped base of the solar cell. This may happen if there

3.1. RESULTS: FORWARD I–V CHARACTERISTICS

are holes in the n^+ emitter and a silver grid line is printed over this area and fired through the ARC layer. Furthermore, Schottky-type shunts may happen if the front contact is fired through the emitter, which is probable if the firing conditions (see step 6 in Figure 1.1) are wrong. At alkaline texturized solar cells the tips of the random pyramids may be truncated. If this happens before the front contact is printed on the solar cell the silver contact has a direct contact to the p-doped base and therefore causes a Schottky contact.

In general Schottky-type shunts do not show a clear Schottky-type rectifying $I-V$ characteristic like a Schottky diode. For a good rectifying Schottky contact clean metallization conditions are necessary. However, such conditions are not given in solar cell production processes. Hence, only weak rectifying behavior of Schottky-type shunts is expected. Mostly such shunts show also weak signal if the solar cell is reverse biased. The $I-V$ characteristic of Schottky-type shunts is somewhat rectifying, but shows also a certain pre-breakdown behavior. The ideality factor of contaminated Schottky contacts is expected to be also not 1, as for ideal Schottky contact, but considerably larger.

3.2 Results: Reverse I–V Characteristics

As shown in chapter 1, section 1.4.3, reverse-biased solar cells show higher current densities than expected from theory. In this section the physical origins of increased reverse current densities in reverse-biased solar cells are described. Furthermore, two possible defect-induced pre-breakdown meachnisms are compared and the physical cause of unexpectedly low breakdown voltages in acidic texturized solar cells will be shown.

However, beforehand a brief motivation is given why detailed knowledge of the physical origins of high reverse currents in solar cells is of high interest for solar cell research and development as well as for solar cell industry.

In solar cell modules often more than 30 cells are connected in one string to sum up the voltages of the solar cells. If one solar cells is shaded by a leaf, a bird, or everything else you may think of, this solar cell will be reverse-biased by the non-shaded cells of the module. Since solar cells show much lower breakdown voltages than expected by theory, the shaded solar cell may break down electrically and very high reverse currents may flow in this solar cell. The high current may lead to so-called hot spots [36, 37], which may damage the cell and therefore the whole module.

In the last years the efforts of investigations concerning the behavior of partially shaded modules, their electrical and thermal properties, and hot spots in modules have increased. However, older publications for example by Hermann [89] and recent publications by García et al. [90, 91], and Munõz et al. [92] focus mainly only on modules and cells but do not have a look at the microscopic origins of the high reverse current. As already mentioned at the end of chapter 1, Bishop et al. [36] and Simo et al. [37] tried to find out more about the origins of the hot spot formation, but they ended up with speculations about the mechanism leading to hot spots due to avalanche breakdown.

Solar cells with high reverse currents are off-specification cells, so they can not be used in modules. Therefore, a deeper insight into the physical origins of pre-breakdown in solar cell may help to identify the responsible mechanism and may help to avoid off-specification cells by improvement of the material quality or the solar cell process. Generally, the detailed physical behavior of large area p-n junctions like solar cells under reverse bias was unknown so far. Therefore, there is a need for fundamental research on reverse-biased solar cells.

The fundamental question is: why do solar cells show a much lower breakdown voltage than expected from theory?

Please note that all pre-breakdown sites shown in this section imaged by LIT techniques are only visible if the solar cells are reverse biased. None of them have a linear (ohmic) fraction. If there are strong ohmic shunts in a solar cell, they usually also dominate under reverse bias.

3.2.1 The Term "Pre-Breakdown"

The breakdown voltage U_b of a Si p-n junction depends on its net charge carrier density. Regarding solar cells as flat large area p-n junctions with a base doping of about 10^{16} cm^{-3}, they should break down at about –60 V. This is illustrated in Figure 3.17, which is obtained from [16], and shows the breakdown voltage of Si for different doping densities. Real multicrystalline-Si solar cells break down at reverse biases significantly lower than –60 V. This is schematically shown in Figure 3.18. The blue dashed curve displays the reverse I–V characteristic corresponding to the calculated breakdown volt-

3.2. RESULTS: REVERSE I–V CHARACTERISTICS

age of –60 V, the red solid curve is a typical real reverse $I-V$ characteristic of a real solar cell. As can be clearly seen in Figure 3.18, real solar cells break down before the theoretically expected breakdown voltage is reached. This is the reason why such breakdown behavior is called "pre-breakdown". In mc-Si solar cells all breakdowns are pre-breakdowns and according to the description in chapter 1 they are devided in three classes: linear, defect-induced pre-breakdown (as will be shown later), and hard pre-breakdown, which is shown in Figure 3.18. Real reverse $I-V$ characteristics may differ

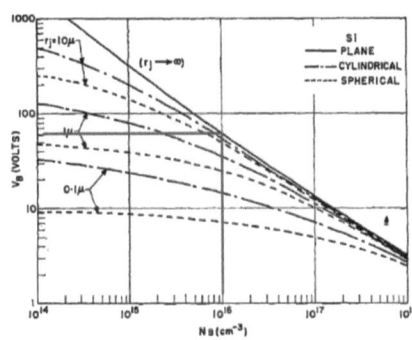

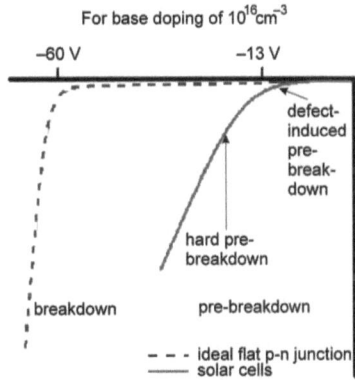

Figure 3.17: Plot of the breakdown voltage U_b (here U_b is called V_b) versus the base doping concentration of Si p-n junctions taken from [16].

Figure 3.18: Breakdown of an ideal flat p-n junction (blue dashed line) in comparision with pre-breakdown of solar cells (red solid line).

significantly. Nevertheless, some characteristic features of reverse $I-V$ characteristics can be defined. In Figure 3.19 the reverse $I-V$ characteristics of an acidically texturized and an alkaline texturized standard industrial solar cell are displayed linearly in a) and semi-logarithmically in b). The $I-V$ characteristics are taken in the dark and at $T = 25$ °C and both solar cells used are free of ohmic shunts. Both solar cells show a linear behavior at low voltages up to approximately –2 V (acidic) and – 3 V (alkaline). The $I-V$ curve of the alkaline texturized solar cell (solid curve) shows a higher slope than the acidic one (red dashed line). This behavior is much more clearly pronounced in the semilogarithmic plot in Figure 3.19b). At moderate reverse voltages from –2 V to –13 V for the acidic solar cell and –3 V to –15 V for the alkaline cell the current increases almost exponentially, here the slope of the acidic cell is slightly higher than that of the alkaline solar cell. At –13 V the reverse current of the acidic solar cell increases strongly, the same behavior is observed for the alkaline cell at about –15 V. This range is called hard pre-breakdown and is characterized by a sudden increase of the reverse current. Alkaline texturized solar cells have a hard pre-breakdown at larger reverse voltages than acidically texturized solar cells. The reason for that behavior will be shown later in this section.

The kinks in the $I-V$ characteristics which occur if hard pre-breakdown takes place are better visible in the linear $I-V$ diagram. On the other hand the semi-logarithmical diagram reveals that the reverse current increases almost exponentially. By extrapolating the steepest part of the linearly plotted $I-V$ characteristic in the hard pre-breakdown range down to zero ampere, the "breakdown voltage" of the solar cell can be estimated. As already mentioned in chapter 1, there is no clear definition of the breakdown voltage. In particular this is true for solar cells.

CHAPTER 3. I–V CHARACTERISTICS OF SOLAR CELLS

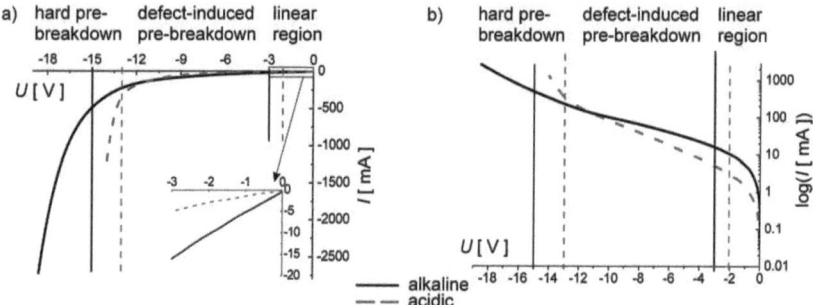

Figure 3.19: I–V characteristics of an acidic texturized solar cell (red dashed curve) and an alkaline texturized solar cell (black solid curve). In a) the I–V characteristics are plotted linear, whereas the characteristics in b) are plotted semi-logarithmically.

Temperature Dependent LIT Measurements

In the next subsections detailed descriptions and discussions of the different pre-breakdown ranges are given. For that purpose temperature dependent, reverse bias DLIT-, ILIT-, and I–V measurements on an acidically texturized solar cell (here called "solar cell 1") are performed. The range for the applied reverse voltage was from –0.5 V to –14 V and the temperature ranged from 25 °C to 80 °C. Using these data a full data set of the physical parameters of solar cell 1 is generated by applying the newly developed methods (TC-DLIT, slope-DLIT, MF-ILIT; see section 2.1.2). The data set consists of:

Figure A.1: a 2-dimensional (2D) set of current densitiy images calculated by equation (2.11) from the measured $S^{-90°}$ images to display the current density J,

Figure A.2: a 2D set of TC-DLIT images calculated by equation (2.12) from the current density images to diplay the TC,

Figure A.3: a 2D set of slope-DLIT images calculated by equation (2.13) from the current density images to display the slope of the reverse current,

Figure A.4: a 2D set of MF-ILIT images calculated by equation (2.17) from temperature dependent $S^{-90°}$ ILIT images to display the MF of the solar cell.

All image sets are shown in the appendix A, since they are too large to be included in the text.
  In the next subsections some selected images from these sets are shown for explanation. However, the sets themselves reveal many more trends and changes of the respective physical parameters and correlations between them. The reader is referred to the corresponding set if necessary. For better orientation one linear pre-breakdown site in the 2-dimensional image sets in the Figures A.1 to A.4 and in all Figures which appear in the text is marked exemplarily by a dashed circle. In addition to the linear pre-breakdown one defect-induced pre-breakdown site in the 2-dimensional images sets and in the Figures in the text is marked by a solid circle. Also exemplarily one hard pre-breakdown site is marked by a dotted circle.

3.2.2 Early Pre-Breakdown in Solar Cells: Linear Region

Early pre-breakdown sites are point-like pre-breakdown sites, which are visible already at very low voltages. In Figure 3.20 an $S^{-90°}$ image taken at –4 V and 25 °C of solar cell 1 is shown in a), as well as a TC-DLIT image in b), and a slope-DLIT image in c). The DLIT signal of the early pre-breakdown

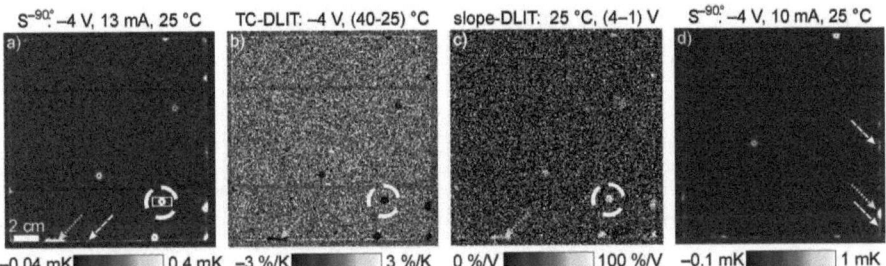

Figure 3.20: a) $S^{-90°}$ image, b) the TC-DLIT image, c) slope-DLIT image of solar cell 1. The dashed circle marks the position of an early pre-breakdown site. In d) the corresponding $S^{-90°}$ image of an adjacent cell of solar cell 1 shows early pre-breakdown sites at different positions compared to a).

sites is strongly temperature dependent, which can be seen in in the current density images Figure A.2 (e.g. at –4 V). The early pre-breakdown sites show a negative TC, which becomes weaker at higher temperatures (cf. Figure A.2 at –4 V or –7 V) and they have very weak slopes of the reverse current (cf. Figure A.3). Early pre-breakdown sites do not show an MF-ILIT signal.

Interestingly, the –90° images at these sites show a dark spot in the middle, which vanishes at higher reverse bias (see Figure A.1). This observation points to the fact that the current through this pre-beakdown site does not flow continously but reduces with time. This may be a result of the strong negative TC of these sites which will be discussed below. Some early pre-breakdown sites appear at slightly higher reverse voltages than –4 V. This is shown in Figure A.1 in the images taken at –7 V, the early pre-breakdowns which appear somewhere between –4 V and –7 V are marked by solid arrows.

The physical origin of early pre-breakdown is not clear yet. In Figure 3.20d) a DLIT image of an adjacent solar cell to solar cell 1 does not show an early breakdown at the same position, as one may expect if early breakdown sites are correlated to extended defects in the multicrystalline Si block. The early pre-breakdown site of solar cell 1 was cut out of the solar cell by a dicing saw (rectangle in Figure 3.20a) and was investigated in detail. In Figure 3.21a) a high resolution DLIT image, the DLIT image overlayed by the topography image, and a reverse EL image of that sample are shown. Furthermore an SEM image of the sample is given in Figure 3.21b). The bright spot in the images in Figure 3.21a) is the early pre-breakdown site (see solid arrow in the Figure) marked by an ink dot to recover the pre-breakdown site for further investigations. Also a reverse bias EL signal can be observed there. Unfortunately, nothing special can be found at the considered position on the surface neither by light microscopy nor by SEM, as can be seen in Figure 3.21b) One may have a closer look, but it is hard to prepare a sample without knowing the exact position, hence the origin of early pre-breakdown is still unclear. In Figure 3.21c) the reverse $I-V$ characteristic of the early pre-breakdown measured at the cut-out sample is shown. It reveals a linear behavior at low reverse voltages and a steep increase of the current at higher reverse voltages. By DLIT imaging (not shown here) it was demonstrated that the linear behavior is not caused by shunting due to defects introduced by sawing out the sample. No additinal shunts could be measured by DLIT. Generally a couple of these early

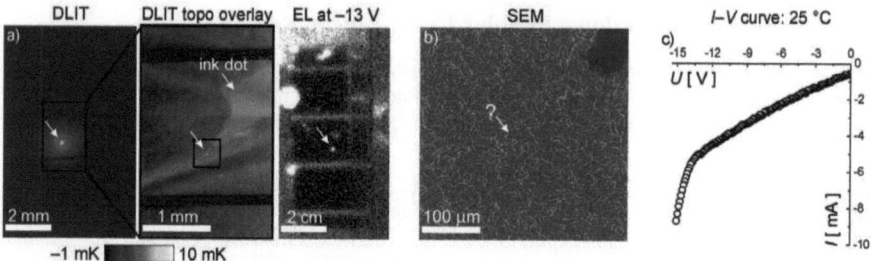

Figure 3.21: a) shows an amplitude DLIT image of the sample cutted out from solar cell 1 (see rectangle in Figure 3.20a) tilted by 90°, an overlay of the amplitude image with the topo image and the corresponding EL image of that sample. In b) an SEM image is shown, unfortunately nothing unusual could be found. In c) the $I-V$ characteristic of that sample is shown. (The arrow marks the position of the early pre-breakdown site.)

pre-breakdowns are observed in solar cells. By summing up the reverse current which is dragged through several early pre-breakdown sites, the linear behavior of $I-V$ characteristics of mc-Si solar cells at low reverse biases, as shown in the inset of Figure 3.19a), can be explained. The steep increase at about -14 V is probably due to the onset of other breakdown sites, which can seen in reverse bias EL (see EL image in Figure 3.21).

As can be seen in Figure 3.20a) and d) (dashed arrows) also the edges of solar cells contribute to the reverse current at low voltages. Hopping conductivity is expected here to be the electron transport mechanism, and this current is assumed to be linear as well [86]. However, hopping conductivity shows a positive TC, so it cannot be responsible for the early pre-beakdown sites. Note that the bright spots at the edges marked by the yellow dotted arrows in Figure 3.20a) and d) are also early pre-breakdown sites which is proven by the fact that their TC-DLIT signal is also negative and they show the same slope as the early pre-breakdown sites in the area.

Since the cause of early pre-breakdown is not known yet, one can only speculate about the origins of their electrical behavior. Early pre-breakdown sites show a negative TC so one may conclude that avalanche, possibly driven by defects, takes place, but this would lead rather to a superlinear $I-V$ characteristic than to a linear one. Another argument against avalanche is that no avalanche multiplication is measurable by MF-ILIT at early pre-breakdown sites. Furthermore, the possible defect then has to be a very local defect, otherwise the same early pre-breakdown would appear in the adjacent cell. Due to their low slope, these early pre-breakdown sites are negligible at high reverse bias, hence they are not very dangerous.

3.2.3 Pre-Breakdown in Solar Cells: Defect-Induced Breakdown

Now a description of the behavior of pre-breakdown sites occurring at moderate voltages in mc-Si solar cells is given. Regarding the $I-V$ characteristics in Figure 3.19 the current in the pre-breakdown region increases exponentially. In Figure 3.22 temperature dependent $I-V$ characteristics for 25 °C, 40 °C, 60 °C, and 80 °C of solar cell 1 are displayed. The pre-breakdown region from about -3 V to -13 V is characterized by an increased current with increasing temperature. In Figure 3.23 one of the corresponding current density images, one TC-DLIT, one slope-DLIT, and the forward EL image of solar cell 1 are shown, respectively. The areas of increased reverse current in Figure 3.23 show a negative temperature coefficient, but their TC is not as negative as the TC of the early pre-breakdowns

3.2. RESULTS: REVERSE I–V CHARACTERISTICS

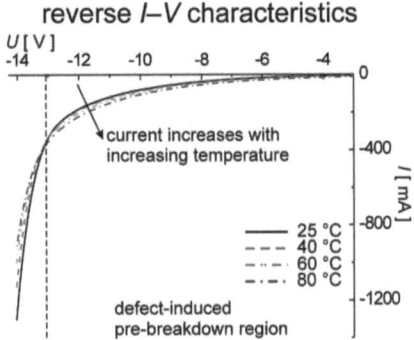

Figure 3.22: Temperature dependent $I-V$ characteristics of solar cell 1.

in that temperature and voltage range. As one can see in Figure A.2 the negative TC becomes weaker for increasing temperatures and increasing voltages, respectively. Moreover, the TC reaches values of about zero or even becomes slightly positive at higher temperatures as can be seen in the image −11.5 V vs. $(80-60)$ °C in Figure A.2. For voltages < -13 V and T_{mid} above 50 °C [$(60-40)$ °C] the TC of these pre-breakdown sites is always about zero %/K. Figure 3.23 reveals that the slope of the

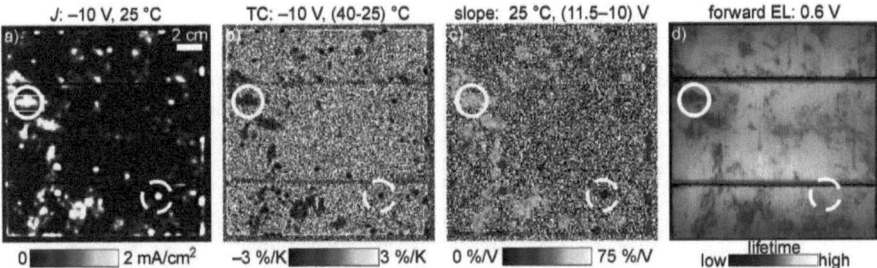

Figure 3.23: a) current density image, b) TC-DLIT image, c) slope-DLIT image taken in the pre-breakdown region. In d) a forward bias EL image of solar cell 1 is shown.

pre-breakdown current is higher compared to the slope of the early pre-breakdown sites. The slope is not temperature dependent, which can be seen in Figure A.3.

For a closer look on a pre-breakdown site, the piece marked by the white rectangle in Figure 3.23a) was cut out of the solar cell 1. In Figure 3.24a) two reverse bias EL images taken at −10 V and −13 V, respectively, are shown. The positions of the EL signals of the pre-breakdown sites correspond to the DLIT signals as can be seen in the EL DLIT overlay image in Figure 3.24b). The EL signal comes from a lot of point-like sources and originates from defects causing recombination. The $I-V$ characteristic in Figure 3.24c) shows a nearly exponential behavior. Where does the current at pre-breakdown sites come from? By comparing the LIT images with the forward EL image in Figure 3.23, a correlation of regions with low lifetimes and the pre-breakdown sites is observed. It is well known that regions of low lifetime in mc-Si solar cells are regions of recombination-active defects. These

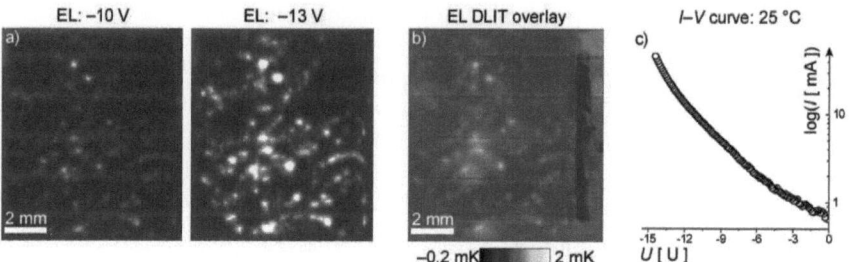

Figure 3.24: a) reverse EL images, b) overlay of the EL image at −10 V with the corresponding amplitude DLIT image, c) I–V characteristic of the of the cut-out sample.

defects can be grain boundaries, dislocation networks, or other crystal defects which are decorated with foreign atoms. The image comparision shows that regions where recombination-active defects are observed break down at moderate reverse voltages. What is the physics behind that pre-breakdown mechanism?

The pre-breakdown behavior can not be explained by just one of the well known breakdown mechanisms, internal field emission or avalanche breakdown. Indeed, pre-breakdown sites show a negative temperature coefficient, which is characteristic for avalanche breakdown. However, the pre-breakdown sites do show only a very weak avalanche multiplication (cf. Figure A.4) only at high voltages and low temperatures. No avalanche multiplication is observed at higher temperatures and at higher voltages, respectively. At higher voltages and temperatures the pre-breakdown sites show a TC close to zero or even a very weakly positive TC. So maybe at higher temperatures some kind of internal field emission takes place, whereas at lower temperatures avalanche breakdown with a low multiplication factor occurs. Hence, it is assumed that more than one breakdown mechanism or better to say a mixture of both internal field effect and avalanche breakdown occur. The pre-breakdown is caused by the recombination-active defects which are found at the pre-breakdown sites. Note, however, that internal field emission according to Figure 1.8 is very improbable at doping concentrations as low as 10^{16} cm^{-3}. Moreover, Lausch et al. [93] have shown that this type of breakdown also appears on flat p-n junctions where avalanche breakdown according to Figure 1.7 is also improbable at such low voltages. Therefore a breakdown mechanism involving defect states appears to be the most probable mechanism for this type of breakdown.

Proposals for Defect-Induced Pre-Breakdown Mechanisms

Defects in semiconductors such as crystal defects or impurities provide energy levels in the band gap. These defect levels are non-radiative recombination paths and therefore decrease the EL signal in forward biased solar cells. Such defects may also act as recombination paths or as traps for electrons and holes in reverse-biased p-n junctions leading to increased leakage currents already at lower reverse voltages than expected from theory. Hence, defect levels within the p-n junction may cause a reverse current which is much higer than the expected saturation current.

Defect levels in the p-n junction of a semiconductor may lead to trap-assisted tunneling (TAT) [20], and one may consider TAT as a possible breakdown mechanism at pre-breakdown sites. In Figure 3.25 trap-assisted tunneling is described schematically. Here the position of the defect level E_{defect} is considered to be in the middle of the gap, where it is most effective. Therefore, the tunnel barrier is

3.2. RESULTS: REVERSE I–V CHARACTERISTICS

lowered significantly compared to band-to-band tunneling (Zener tunneling). This may explain the enhanced reverse current at low reverse voltages at the defects. Nevertheless, the temperature coefficient of the reverse current of trap-assisted tunneling is still positive [94]. This is at least in contradiction to the measured TC for pre-breakdown sites at low temperatures. However, at higher temperatures and also higher voltages (see Figure A.4), respectively, the TC of the pre-breakdown changes to values slightly above zero. It might be possible that trap-assisted tunneling becomes the dominant reverse current channel for higher temperatures. Still the question remains: What is the cause of the negative

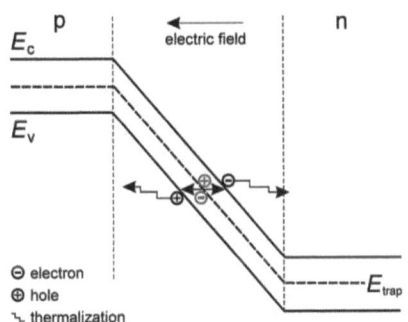 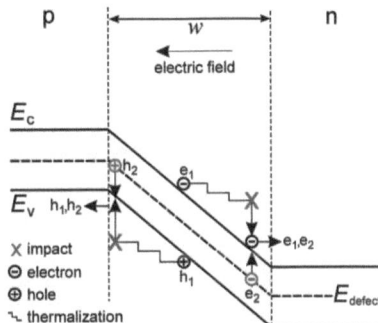

Figure 3.25: Trap-assisted tunneling schematically.   **Figure 3.26:** Trap-assisted avalanche schematically.

TC of the pre-breakdown sites at lower temperatures and voltages? The presence of defects within the p-n junction may also lead to another kind of trap assisted charge carrier transport, which may be called trap-assisted avalanche (TAA). In Figure 3.26 TAA is shown schematically. Here the electron "e_1" is accelerated by the electric field within the depletion layer width w. If the defect level E_{defect} is occupied by an electron "e_2", the accelerated electron "e_1" may ionize the defect state by exciting the electron "e_2" to the conduction band leaving an unoccupied defect level. TAA works similarily also for holes, as shown also in Figure 3.26. It is also possible that free electrons are ionizing bond holes and vice versa. If an electron is emitted by a TAA process, the level is occupied by a hole, which may be ionized by a following TAA process. Whether the defect state is mainly occupied by a hole or an electron is depending on the type of the defect, i.e. the capture cross section for electron and holes, and on the average electron and hole concentration in the depletion layer. The necessary succesive electron and hole emission may be the reason for the fact that the multiplication factor for this type of avalanche is smaller than for classical avalanche. Trap-assisted avalanche will show a negative temperature coefficient of the current. The reason is that at higher temperatures TAA becomes improbable because of the increased phonon scattering of the accelerated charge carriers. Hence, the probability that an accelerated charge carrier can gain enough energy to ionize a defect state becomes smaller with increasing temperatures.

The Measurements in Detail

The slope of the reverse current at defect-induced pre-breakdown sites is constant at a value of about 60 %/V for all temperatures and voltages, which can be seen in Figure A.3.

However, if one has a look at the current density images taken at –7 V in Figure 3.27a) (taken from Figure A.1), the pre-breakdown sites marked by the solid circles vanish at temperatures higher than

40 °C. The TC of these pre-breakdown sites is negative at –7 V and $T_{mid} = 32.5$ °C as can be seen in Figure 3.27b) (taken from Figure A.2). The TC signal becomes weaker at $T_{mid} = 50$ °C and vanishes completely at $T_{mid} = 70$ °C. This behavior is typical for avalanche breakdown. However, there was no avalanche multiplication measureable at such low voltages. Regarding the same current density

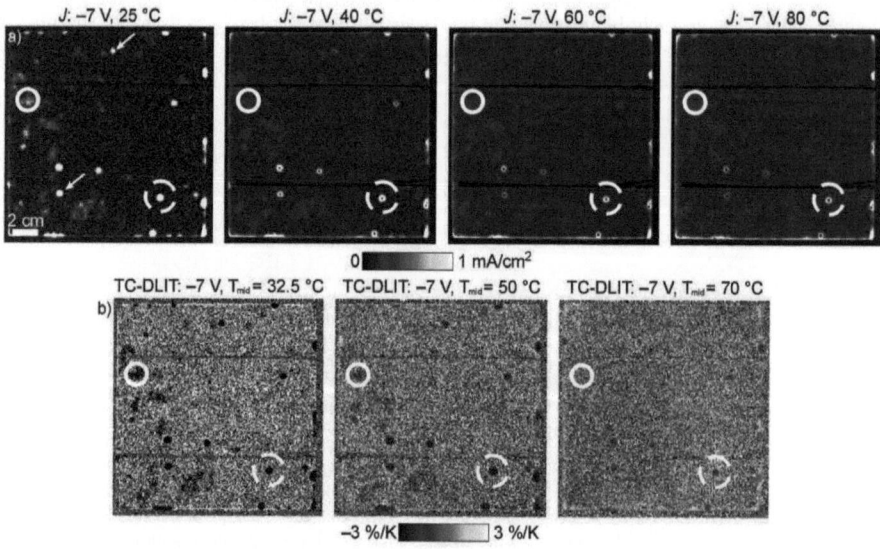

Figure 3.27: a) current density images at –7 V, and b) TC-DLIT images at –7 V of solar cell 1, respectively.

data for –10 V in Figure 3.28a) (taken from Figure A.1) one can see that the pre-breakdown sites marked by the solid circles do not vanish at higher temperatures. The current density decreases only very slightly or is even constant with higher temperatures. This is proven by the fact that the TC of the pre-breakdown sites becomes less negative or slightly positive as can be seen in Figure 3.28b) (taken form Figure A.4). A positive TC is a clear hint for internal field emission mechanism for breakdown. Regarding the measured behavior of defect-induced pre-breakdown sites it is assumed that a mixture of both defect-induced breakdown mechanisms, trap-assisted avalanche and trap-assisted tunneling, beeing responsible for defect-induced pre-breakdowns. Both mechanisms seem to be active at the same time, but at temperatures below 40 °C and at voltages > -10 V trap-assisted avalanche is assumed to be the dominant mechanism showing a negative TC. Obviously, trap-assisted avalanche is suppressed immediately if the temperatures reach 40 °C, and trap-assisted tunneling is then assumed to be the dominant breakdown mechanism.

The weakness of this hypothesis is the fact that at such low voltages, where trap-assisted avalanche is assumed to be the breakdown mechanism, no MF signal could be measured directly by MF-ILIT. However, a very weak avalanche signal is observed at the position of the pre-breakdown sites at reverse biases < -13 V, as can be seen in the multiplication factor images at (14.5/13) V and (15/13) V (all taken at 25 °C) in Figure A.4. It has been discussed above that the reason for this low MF may be the necessity for subsequent electron and hole emission for trap-assisted avalanche. Much more effort has to be spent to reveal the types of the involved defects, to get more information on the

3.2. RESULTS: REVERSE I–V CHARACTERISTICS

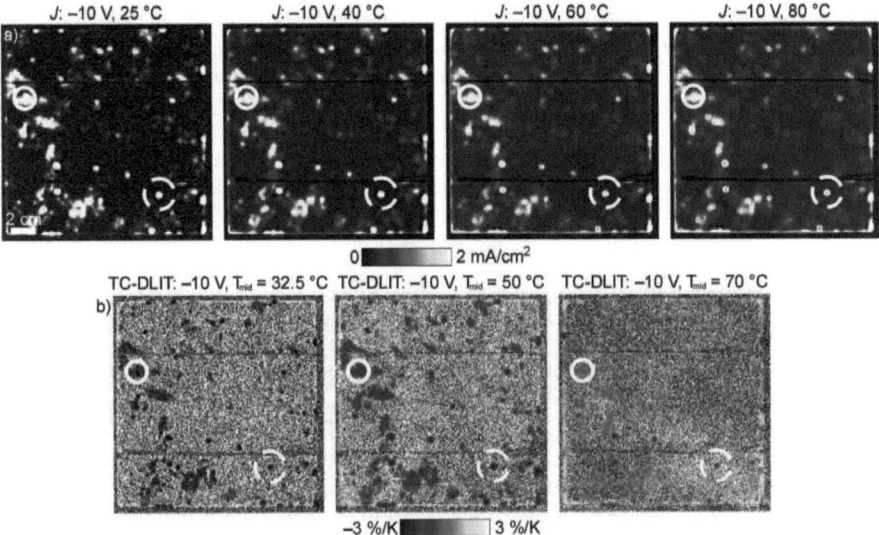

Figure 3.28: a) current density images at −10 V, and b) TC-DLIT images at −10 V of solar cell 1, respectively.

exact breakdown mechanism, and to prove or disprove the hypothesis made here. A more detailed investigation of the defects will be one of the most interesting tasks for future work.

Homogenous Reverse Current

By having a closer look at the I–V characteristic in the range of the defect-induced pre-breakdown, i.e. about −3 V to −13 V, and comparing it with the TC-DLIT images, a contradiction is observed. In

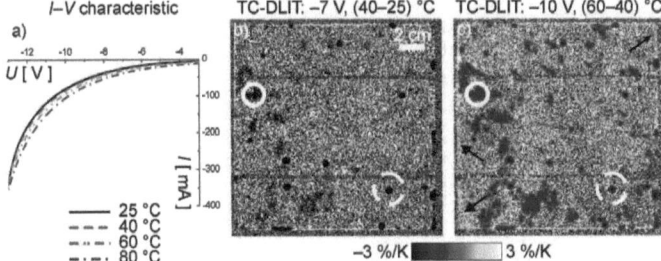

Figure 3.29: a) shows the I–V characteristic of solar cell 1 in the pre-breakdown range, in b) and c) corresponding TC-DLIT images are shown.

Figure 3.29a) the respective region of the temperature dependent I–V characteristics is shown. The current increases clearly with increasing temperature. However, in Figure 3.29b) and c) TC-DLIT images from the corresponding voltage range show a lot of sites which have a negative TC. Only some pre-breakdown sites at the edge of the cell have a positive TC (marked by solid arrows in Figure 3.29c), but there are much fewer sites showing positive TC than sites showing negative TC. This is, at the first sight, in contradiction to the measured $I-V$ characteristics. Where does the positive TC of the latter come from, when almost all pre-breakdown site show a zero or negative TC? In Figure 3.30 the current density images of solar cell 1 taken at −10 V and four temperatures, respectively, are shown. Here the current density images are scaled at lower mA/cm^2 values than in Figure A.1. In the areas between the pre-breakdown sites a clear increase of the current density with increasing temperature is observed.

The current between the pre-breakdown sites does flow essentially homgenously in the whole area of the solar cell and is very weak. Hence, the signal is very noisy. By averaging the current density of the square marked in the current density images of Figure 3.30 the current density signal was improved and was used for analysis. This marked area does not contain any pre-breakdown site. In Figure 3.31a) the average current density at a constant voltage (here −10 V) depending on temperature is plotted. The average current density increases with temperature but starts to saturate at 60 °C. In

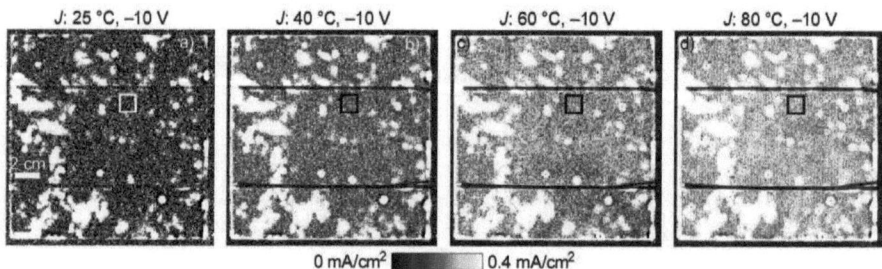

Figure 3.30: Current density images of solar cell 1 at −10 V for different temperatures.

Figure 3.31b) the dependency of the average current density on the voltage for the different temperatures is shown. The higher the temperature the higher is the current density, but the shape of the curves is similar. The average current density increases slightly superlinearly from −0.5 V to about −10 V. At voltages < -10 V the average current density saturates or even becomes smaller when the voltage is further increased. The strong fluctuations of the average current densities are due to the very noisy LIT signal. As one can see in the TC-DLIT images for example in Figures 3.27b) and 3.28b) the noisy TC-DLIT signal between the pre-breakdown sites is slightly brighter for the lower $T_{mid} = 32.5$ °C than for $T_{mid} = 50$ °C and $T_{mid} = 70$ °C. Hence, the noisy signal should be taken into account for evaluating the distribution of temperature coefficient in solar cells. For a better TC-DLIT analysis the signal-to-noise ratio for the measurement of a solar cell processed from a wafer adjacent to solar cell 1 was improved by binning 3×3 pixels of the current density images and afterwards calculating the TC-DLIT images. In Figure 3.32 the TC-DLIT images calculated from the binned current density images are shown for all T_{mid} at −10 V. In contradiction to the TC-DLIT images shown in 3.28b) in Figure 3.32 areas of clearly positive TC are observed. The TC becomes less positive with increasing temperature, but as already mentioned above also the sites of clearly negative TC become less negative or even slightly positive, so the contribution of sites of negative TC becomes smaller (cf. Figure 3.28b). The scaling for the TC-DLIT images in Figure 3.32 is different.

3.2. RESULTS: REVERSE I–V CHARACTERISTICS

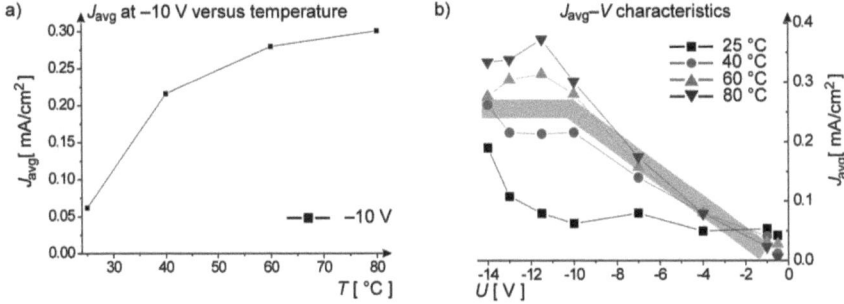

Figure 3.31: In a) the average current density at –10 V depending on temperature taken from the square in Figure 3.30 is plotted. b) shows the average current density over the whole reverse voltage range for different temperatures. The thick gray line in the background indicates the overall shape of the curves.

Note that the TC-DLIT images in Figure 3.32 also reveal that the structure of sites of positive and negative TC are nearly equal for solar cells made from adjacent wafers of an mc-Si block. This is a direct proof that defect-induced pre-breakdown sites are caused by extended defect structures like e.g. electrically active grain boundaries in the mc-Si block as mentioned above.

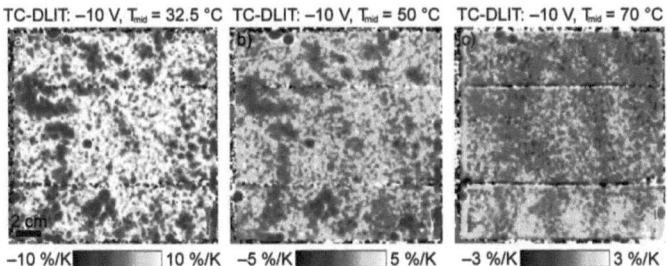

Figure 3.32: TC-DLIT images calculated after binning of 3 × 3 pixels of the current density images. (Note that these images are taken at a solar cell made from a wafer adjacent to the wafer used for solar cell 1, please note also the different scaling of the images.)

Overall I–V Characteristics in Defect-Induced Pre-Breakdown Region

The reverse current of a solar cell in the defect-induced pre-breakdown region is determined by the exponential current flowing due to defect-induced pre-breakdown and the slightly superlinear current which flows through the whole cell area. The latter current dominates the temperature dependence of the entire $I-V$ characteristic in the pre-breakdown region. The overall current in the pre-breakdown region is the sum of the current flowing through the defect-induced beakdown sites and the homogenous current. At higher temperatures the homogenous current in the area saturates, which can also be seen in the slope images in Figure A.3. Since the current due to defect-induced pre-breakdown is exponential and the homogenous current is slightly superlinear, the overall $I-V$ characteristic of the pre-beakdown region of a solar cell is approximately exponential, too. Until now the nature of the

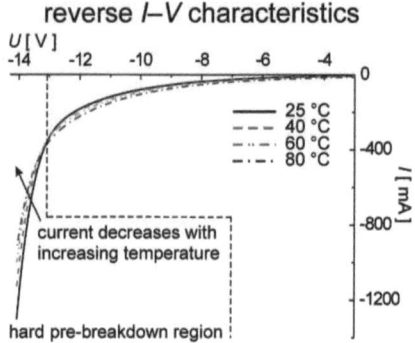

Figure 3.33: Temperature $I-V$ characteristics of solar cell 1.

homogenous current is unknown. With the described properties it is very unlikely that it can be explained by a saturation current of the two-diode model (see section 1.2.2). In contradiction to Figure 3.31, such a current should saturate at < 1 V and should depend exponentially on the temperature.

3.2.4 Hard Pre-Breakdown: Avalanche Breakdown in Solar Cells

The last region of the $I-V$ characteristic of an mc-Si solar cell shows a very steep increase of the reverse current at voltages < -13 V, as can be seen at the $I-V$ characteristic in Figure 3.33. Since the current increase is very sudden this pre-breakdown region is called hard pre-breakdown. The hard pre-breakdown region is most interesting because of its sudden occurrence and the high slope of the reverse current. In Figure 3.34 $S^{-90°}$ images of solar cell 1 at several higher reverse voltages at constant temperature, and a forward EL image of solar cell 1 are shown. The dotted circles and ellipses mark sites of hard pre-breakdown. In the $S^{-90°}$ images taken at −11.5 V, Figure 3.34a), no signal can be observed at the hard pre-breakdown sites, but they suddenly show up at −13 V in b) and are even more pronounced at −14 V in c). The overall current of the solar cell from −11.5 V (155 mA) to −13 V (371 mA) is more than doubled, and at −14 V (1368 mA) it is more than 3.5 times higher than at −13 V. Figure 3.34 reveals that this high current mostly flows through some localized sites in the cell. It is true that the current at the early pre-breakdown sites and at the defect-induced pre-breakdown sites increases also. However, the slope of the reverse current at these sites is small and the contribution to the increasing reverse current comes mainly from a few hard pre-breakdown sites. The hard-breakdowns sites do not correspond to sites of recombination-acitve defects in the solar cell as can be seen by comparing Figures 3.34b) and c) with Figure 3.34d). In the forward EL image in d) the positions of the hard pre-breakdown are marked, but no recombinative defects can be observed here. Hence, defect-induced breakdown mechanism can not be taken into account as the breakdown mechanism for the hard pre-breakdown sites.

In Figure 3.35 a TC-DLIT, a slope-DLIT, and an MF-ILIT image of solar cell 1 in the region of hard pre-breakdown are shown. Additionally, in Figure 3.35d) the $I-V$ characteristic measured at the cut-out part marked by the white rectangle in Figure 3.34c) is given. As one can see in Figure 3.35 the hard pre-breakdown sites show a clearly negative TC, a high slope of the reverse current, and a multiplication factor greater than one.

3.2. RESULTS: REVERSE I–V CHARACTERISTICS

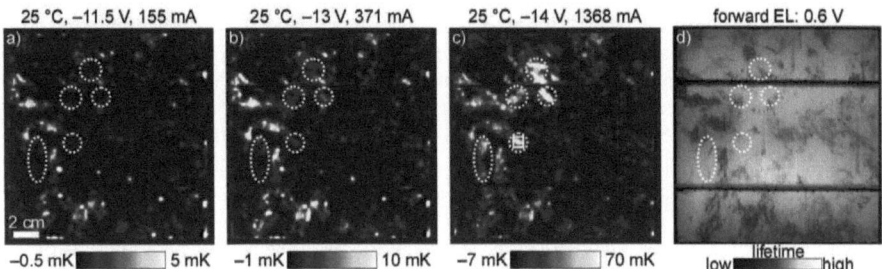

Figure 3.34: In a) to c) $S^{-90°}$ images of solar cell 1 at higher reverse biases are shown. In d) the corresponding forward EL image is shown.

The current density images taken at −13 V and −14 V in Figure A.1 reveal that the hard pre-breakdown becomes slightly weaker with increasing temperature. According to this the TC of the hard pre-breakdowns becomes less negative with increasing temperature (see Figure A.2) but it keeps being negative over all temperatures and voltages. The slope of the reverse current also decreases slightly, which can be seen in Figure A.3. Figure A.4 reveals that the MF decreases strongly with rising temperature. Both the I–V characteristic of the whole solar cell in Figure 3.33 and the I–V characteristic of the cut-out piece which only contains hard pre-breakdown sites (see 3.35d) clearly reveal a breakdown voltage $U_b \approx -13$ V. In Figure 3.35d) it is shown that the I–V curve of the small piece of solar cell shows a very good reverse I–V characteristic, i.e. it shows a very small reverse

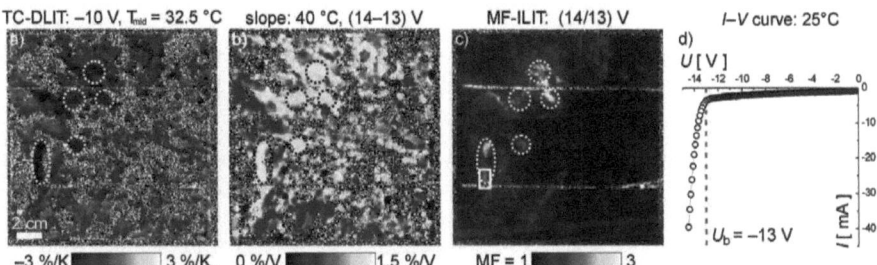

Figure 3.35: a), b), and c) show the TC-DLIT image, slope-DLIT image and MF-ILIT image of solar cell 1 in the hard pre-beakdown region, respectively. In d) the I–V characteristic of the cut-out piece (white solid square in Figure 3.34c) of the solar cell is shown.

current without any early- or defect-induced pre-breakdown until the reverse voltage reaches −13 V. Then the current increases suddenly, this is the hard breakdown. By reaching −13 V the overall I–V characteristic changes its temperature dependency from postive to negative (see Figure 3.33). In the hard pre-breakdown range the overall I–V characteristic is clearly dominated by the hard pre-breakdown.

Since a large fraction of this reverse current flows only through some small spots in the cell, these spots get hot and therefore are called hot spots. They may, in the worst case, damage the solar cell or the whole module. The stable negative TC, the high slope of the reverse current, and the directly measurable charge carrier multiplication lead to the conclusion that avalanche is the responsible mechanism for hard pre-breakdown. Where does the breakdown come from? At hard pre-breakdown sites no extended recombination-active defect could be observed (cf. Figure 3.34), therefore no defect-induced breakdown mechanism can be expected. Furthermore, U_b of −13 V is much lower than the theoretically predicted breakdown voltage of a flat p-n junction. Therefore, no avalanche breakdown at a flat p-n junction can be expected to be responsible for the hard pre-breakdown.

Physical Origin of the Hard Pre-Breakdown

Because of the high interest for solar cell research and solar cell industry, the main results of this subsection have already been published recently in [95]. To reveal the physical origin of the hard pre-breakdown sites, the sample at the position of the white rectangle in Figure 3.35c) showing hard pre-breakdown was cut out of solar cell 1. The $S^{-90°}$ DLIT image and the reverse EL image of that sample as well as an overlay of the latter images are shown in Figure 3.36a) to c). In Figure 3.36d) the I–V characteristic of that sample is shown. As can be seen in Figure 3.36a) the hard pre-breakdown

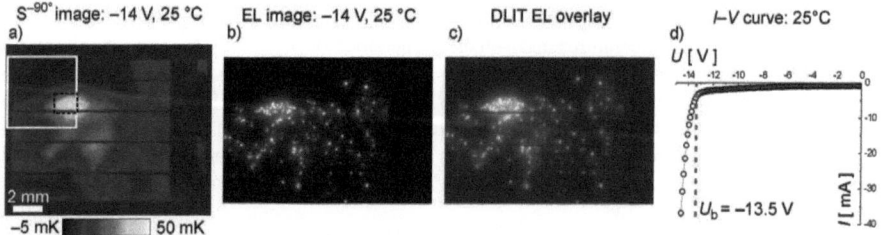

Figure 3.36: In a) the DLIT image of the sample marked in Figure 3.35c), in b) the reverse EL image of that sample, and in c) the overlay of a) and b) are shown. The reverse I–V characteristic of that sample is shown in d).

site is localized at two spots of about 2 mm diameter, respectively. The reverse EL image in Figure 3.36b) reveals that these spots consist of many very small point-like breakdown sites. The outer shape of the EL spot clusters are in agreement with the shape of the hard pre-breakdown sites in the $S^{-90°}$ image, as can be seen in the overlay of the $S^{-90°}$ DLIT image and the EL image in Figure 3.36c). The I–V curve of that sample shows a breakdown voltage of $U_b = -13.5$ V (see 3.36d), which is slightly higher than the breakdown voltage of the hard pre-breakdown sample from Figure 3.35.

The image sequence in Figure 3.37 shows SEM images taken at the hard pre-breakdown site which is marked by the white rectangle in Figure 3.36a). The magnified image in Figure 3.37b) shows microscopic line-shaped structures (marked by the solid arrows), and the high magnification SEM image in c) shows that these lines consist of holes in the solar cell surface (see solid arrow). Furthermore, there are also holes which are not aligned in lines (see dashed arrow in c). The structure which is exemplarily marked in Figure 3.37c) by the dotted arrow is the expected surface structure of an acidically texturized solar cell. To find out if the holes are responsible for hard pre-breakdown lock-in EBIC under reverse bias was performed at the sample. For that purpose the sample was mounted on an SEM/EBIC sample holder which acted also as cooler to avoid drifts of the reverse current due to heating of the sample. For lock-in EBIC a constant reverse voltage in the range of the hard pre-breakdown (−13 V to −15 V) was applied to the sample and lock-in EBIC was performed as described in section 2.3.1. For the lock-in EBIC measurement the circuit in Figure 2.10 with the components $R_1 = 150\ \Omega$, $R_2 = 1$ kΩ, and $C = 63$ μF was used.

3.2. RESULTS: REVERSE I–V CHARACTERISTICS

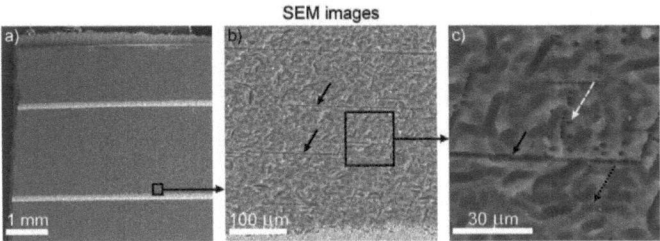

Figure 3.37: SEM images in different maginfications of sample-section marked by the white rectangle in Figure 3.36a).

The aim of this experiment is to generate electron-hole pairs with the electron beam of the SEM by hitting the solar cell sample surface (similar to the IR light used for MF-ILIT). In the case of EBIC an increasing number of electrons due to avalanche leads to an increasing EBIC current and therefore an increasing EBIC signal is excpected at sites where avalanche breakdown takes place. For the experiment an acceleration voltage of the SEM of 30 kV was used. Traditionally, this technique is called microplasma imaging.

The results of the lock-in EBIC experiment at the solar cell sample are shown in Figure 3.38. In 3.38a) the SEM image of an area containing holes arranged in lines is shown. The corresponding lock-in EBIC image is shown in b) and reveals that there is an increased lock-in EBIC signal at the lines. The magnified SEM and lock-in EBIC images presented in Figure 3.38 c) and d) clarify that the increased EBIC signal only comes from the holes in the surface of the sample. In Figure 3.39 a

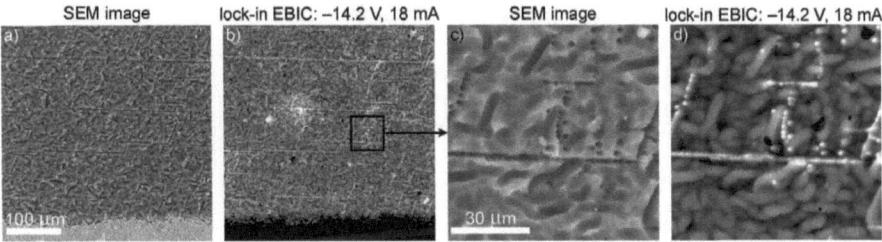

Figure 3.38: In a) an SEM image and in b) the corresponding lock-in EBIC image at a hard pre-beakdown site are shown, the scale bar in a) holds for both, a) and b). c) shows a magnified SEM image and d) the corresponding lock-in EBIC image, the scale bar in c) holds also for d).

lock-in EBIC image sequence taken at the area marked in Figure 3.36a) by the black dotted rectangle is shown. The lock-in EBIC image taken at 0 V (Figure 3.39a) shows a grain boundary (marked by the solid arrow). By increasing the reverse voltage up to −11.1 V there is no significant change in the EBIC contrast observable, but the contrast of the grain boundary becomes weaker. If the reverse voltage reaches −13.5 V, some bright spots appear in the EBIC image, and the signal of the grain boundary vanishes. If the voltage is further increased, some new spots appear and the existing spots change their EBIC contrast. It is hard to evaluate the change in contrast quantitatively, because one has to adjust the brightness and contrast for every new lock-in EBIC image. However, some of the existing spots become relatively weaker (dashed arrow), some vanish (dotted arrow), and some get stronger (dashed dotted arrow). It is important to note that in the upper grain in Figure 3.39 no increased EBIC signals appear. This is nicely show in Figure 3.40a), where no increased EBIC signal can be observed in the upper part of the lock-in EBIC image. In the SEM image the line-shaped structures and holes can only be observed in the lower part. Indeed no holes or line-shaped structures can be found in the upper grain. So, obviously, the appearance of the avalanche sites depends either on grain orientation or the defect content of the grain.

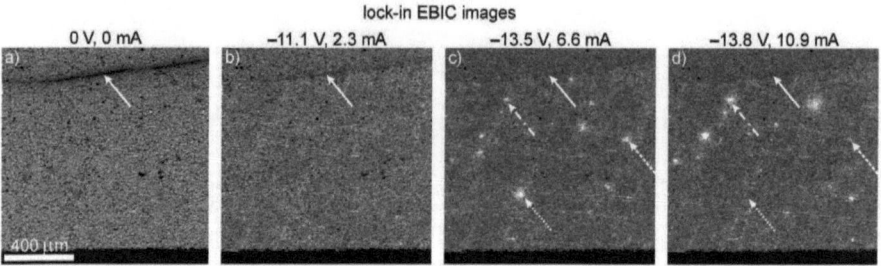

Figure 3.39: Sequence of lock-in EBIC images.

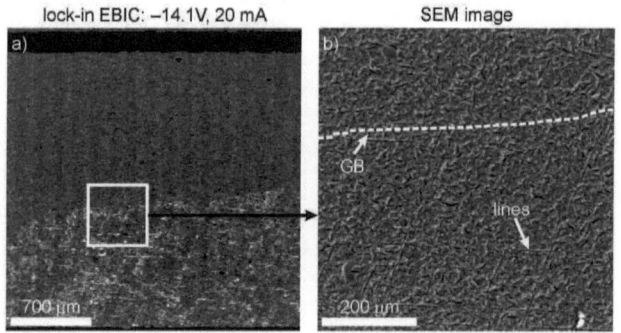

Figure 3.40: A grain with and a grain without lock-in EBIC signal.

Avalanche Breakdown due to Etch Pits

The holes in the front surface of the solar cell are identified to be the origin of the hard pre-breakdown. Why do the holes show a hard avalanche breakdown at about −13.5 V, and how are they formed? In Figure 3.41 SEM images of the surface of the sample marked by the white rectangle in Figure 3.35c) containing hard pre-breakdown sites is shown. The holes are cone-shaped and 1 μm to 3 μm in diameter. The SEM image in 3.41a) was taken with the electron beam being perpendicular to the surface of the sample. As can be seen in Figure 3.41a) the axis of the holes is not perpendicular to the sample surface. If the sample is tilted by 24° with respect to the electron beam (see Figure 3.41b), the axis of the holes and the incoming electron beam are parallel and it was possible to have a look straight into the holes.

3.2. RESULTS: REVERSE I–V CHARACTERISTICS

The holes show a rough surface, and the holes which are aligned along a line show a groove in their middle as can be seen in Figure 3.41c). The holes which are not aligned along a line do not show such grooves, which also can be seen in Figure 3.41c). For revealing the morphology of the holes, FIB cross- and longitudinal sections of the holes are prepared. In Figure 3.41d) such a FIB lamella is shown exemplarily. In Figure 3.42 a detailed investigation of holes showing avalanche breakdown

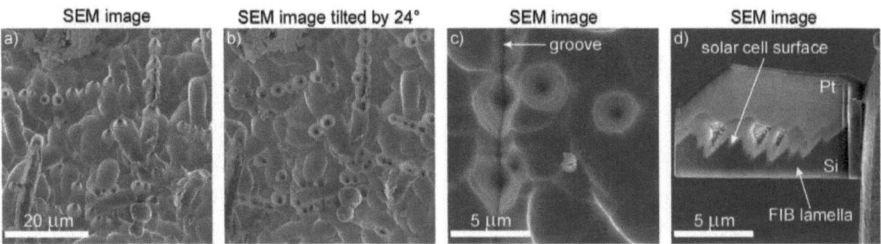

Figure 3.41: a) to c) show SEM images of the surface of a solar cell containing cone shape holes. In d) an SEM image of a FIB lamella is shown. The surface of the solar cell is covered by platinum (Pt) to protect the sample during the focussed ion beam preparation.

is demonstrated. Figure 3.42a) is the lock-in EBIC image taken at −13.7 V. The strongly increased lock-in EBIC signal in the middle of the image has its origin at two holes (see Figure 3.42b). In Figure 3.42c) a detailed SEM image of these two holes reveals that they are formed at two grooves. Exactly at the two holes a FIB lamella was prepared, the position of the FIB lamella is indicated by the dashed line in 3.42c). A TEM image of the FIB lamella is presented in Figure 3.42d). The curvature radius of the tip of the cones is determined to be about 20 nm. At the tip apex some lines are observed, which are crystal defects. Thus the holes are identified to be etch pits formed during

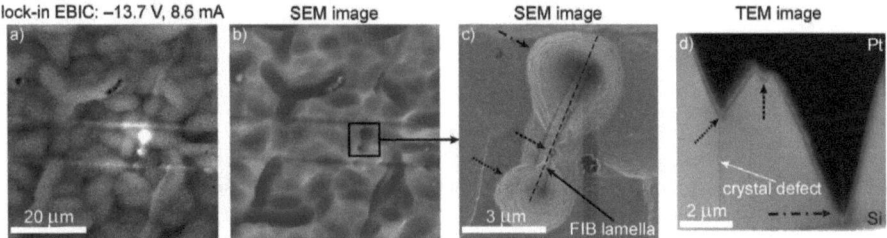

Figure 3.42: Detailed investigation of a cone-shaped hole showing avalanche breakdown. In a) the lock-in EBIC image is shown, the corresponding SEM image in b) reveals that the EBIC signal occurs at the two holes in the middle of the image. In c) a detailed SEM image of the holes is shown, and in d) the TEM image of the cross sectional FIB lamella prepared at the dashed line indicated in Figure 3.42c) is presented. The arrows in c) and d) mark the three holes.

the damage etching/texturization step (cf. step 1 in Figure 1.1). After the etching/texturization step the n^+ layer was diffused into the surface of the wafer, which means that the p-n junction follows the shape of the etched surface. If the solar cell surface contains etch pits, the p-n junction is bent at the etch pits. The bending at the etch pit apex can be assumed to be spherical. In Figure 3.43 the TEM image of one of the holes from Figure 3.42d) is shown. The dashed line indicates the expected shape of the p-n junction schematically. Since the curvature radius of the etch pit apex is only about

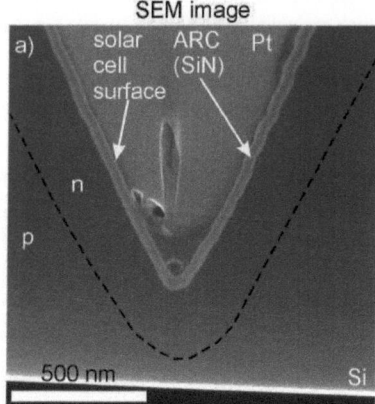
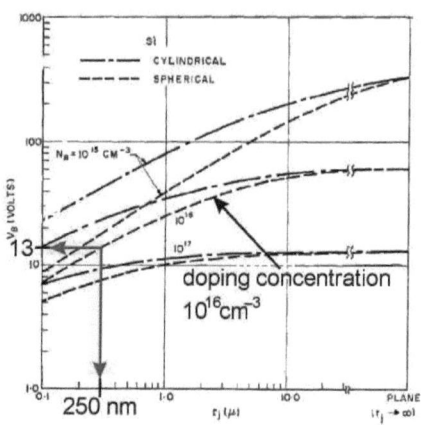

Figure 3.43: TEM image of the cross section of cone-shaped holes showing hard pre-breakdown due to avalanche breakdown.

Figure 3.44: Theoretically predicted breakdown voltage of bent Si p-n junctions, graph taken from [16].

20 nm it is about ten times smaller than the p-n junction width, which is about 250 nm in this type of solar cells. Therefore, the effective curvature radius is determined by the depth of the p-n junction, i.e. 250 nm. Sze and Gibbons have calculated the dependence of the breakdown voltage on the curvature radius of Si p-n junctions at different base doping concentrations [16]. Their curve was confirmed by Speeney and Carey experimentally [96]. In Figure 3.44 the curve calculated by Sze and Gibbons is shown. For a base doping concentration of 10^{16} cm^{-3}, which corresponds to the doping concentration of the p base of solar cells, at a curvature radius of 250 nm the breakdown voltage of a spherically curved Si p-n junction is $U_b \approx -13$ V. This is in perfect agreement with the measurements of the hard pre-breakdown voltage of about -13 V.

Conclusion

By regarding the curvature radius of the tip apex of the cone-shaped holes found in the surface of acidic mc-Si solar cells, the physical mechanism of the hard pre-breakdown sites has been proven to be avalanche breakdown due to field enhancement at the spherically shaped p-n junctions at cone-shaped etch pits. The theoretically [16] and experimentally [96] predicted breakdown voltage of $U_b = -13$ V is in very good agreement with our measurements.

Origin of the Cone-Shaped Etch Pits

How do the etch pits form? Actually the acidic etching solution used for acidic texturization is designed not to produce etch pits at dislocations reaching the surface of the solar cell. At the apex of the etch pits lines are observed as can be seen in the TEM image in Figure 3.42d). These lines are crystallographic defects reaching the surface of the grain. These crystallographic defects are etched during the texturization step of the solar cell and therefore the cone-shaped holes in the surface of the solar cell are formed at these defects. At the positions of the hard pre-breakdown sites, the EL image in Figure 3.34d) reveals that these areas do not show recombination-active defects. From this

3.2. RESULTS: REVERSE I–V CHARACTERISTICS

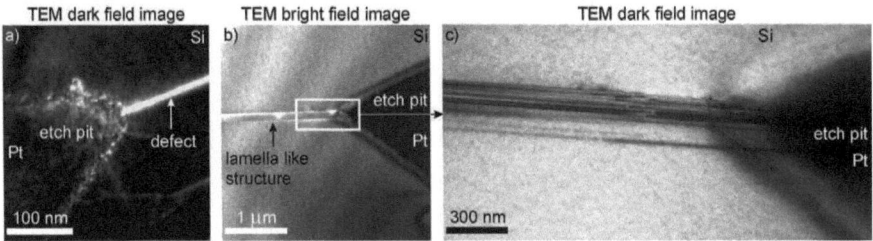

Figure 3.45: TEM images of cone-shaped etch pits in the surface of acidic texturized solar cells. All etch pits show defects at their apex.

it follows that the defects in this areas are not recombinative, but nevertheless their are etched and cone-shaped holes are formed. That means avalanche breakdown occurs at clean crystallographic defects, which can be understood by the fact that clean material is necessary to have sufficient high scattering path length of carriers for enabling the avalanche effect. Note, that also in areas showing recombination the defects may be etched if they reach the surface and form cone-shaped holes. However, at recombination-active defects pre-beakdown takes place already at lower voltages (due to defect-induced mechanisms) than at clean defects.

Grains of mc-Si wafers not showing such crystallographic defects at the surface do not have holes. This is proven by the example shown in Figure 3.40. There, in the upper grain no holes can be found, whereas the lower grain contains a lot of them. In Figure 3.45 TEM images of FIB cross sections of etch pits are shown. The solid arrow in Figure 3.45a) mark a crystal defect pointing exactly to the tip apex of the etch pit. The etch pit marked by the solid arrow in Figure 3.45b) shows a lamella-like structure. However, in Figure 3.45c) a TEM dark-field image of the defect in b) is shown in higher magnifation revealing a very complex structure of the defect, so obviously the defect is not a simple lamella. The only conclusion which can be made until now is that there are possibly several types of defects leading to cone-shaped etch pits. Probably the defects are not single dislocations but rather multiple dislocations. However, the exact type of crystal defects leading to etch pits is not clear yet.

3.2.5 Comparision of Pre-Breakdown I–V Characteristics

The I–V characteristics of the pre-breakdown mechanisms shown in the previous section show very different behavior. In Figure 3.46a) the I–V characteristics of an early, linear pre-breakdown (sample 3), of a defect-induced pre-breakdown (sample 4), of a hard pre-breakdown (sample 2), and of a region of solar cell 1 which do not show any pre-breakdown (sample 1) measured at 25 °C are compared. In order to measure the I–V characteristics only of the certain pre-breakdown site, samples at the postions of the different pre-breakdown sites are cut out of solar cell 1 as carefully as possible with a dicing saw. It was proven that no significant shunts have been generated at the edges of the samples by DLIT imaging after sawing (not shown here). In Figure 3.46b) to d) a TC-DLIT, a slope-DLIT, and an MF-ILIT image are given exemplarily. The positions of the cut-out samples are marked. The I–V measurements on the cut-out samples show very different behavior. The I–V characteristic of the sample without any breakdown (sample 1) shows a flat almost linear I–V characteristic. This is also proven by the TC-DLIT, slope-DLIT, and MF-ILIT image, where no signal can be detected at the position of sample 1.

CHAPTER 3. I–V CHARACTERISTICS OF SOLAR CELLS

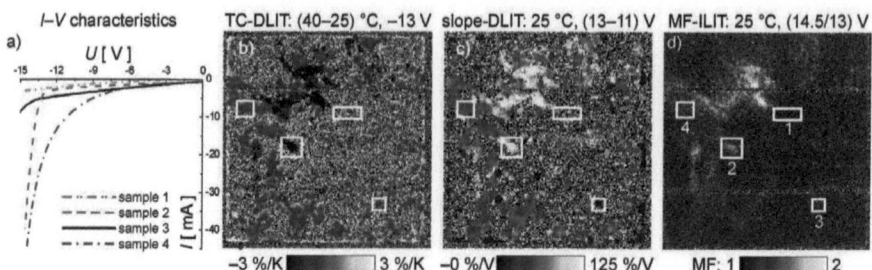

Figure 3.46: In a) the $I-V$ characterisitics of different cut-out samples of solar cell 1 are shown. In b), c), and d) the TC-DLIT, the slope-DLIT, and the MF-ILIT image are shown. The positions of the samples are marked in b), c), and d) and are named in d).

The early pre-breakdown (sample 3) shows also a flat $I-V$ characteristic, but the current is higher than that of sample 1. Furthermore, the current increases suddenly at about −13 V. As demonstrated in Figure 3.46b) and c) the early pre-breakdown shows a negative TC and a very low slope, respectively. However, there is no avalanche MF signal measureable, which could explain the sudden increase of the reverse current at about −13 V. This might be due to the relativly insensitive MF-ILIT method, or one has to assume another breakdown mechanism here which is not know yet.

The defect-induced pre-breakdown (sample 4) shows a superlinear, almost exponential $I-V$ characteristic. The current flowing at defect-induced pre-breakdown sites exceeds the current flowing in early pre-breakdown sites at about −7 V. Then, according to the measured slope shown in Figure 3.46c), the current increases faster. The current of the hard pre-breakdown site (sample 2) at reverse voltages up to −13 V is very low. Already at about −13.5 V the current at the hard pre-breakdown site equals that of the defect-induced pre-breakdown site.

Two very important conclusions can be drawn from these $I-V$ measurements. Firstly, the reverse current at reverse voltages up to −7 V is the sum of the reverse current of all pre-breakdown sites. From −7 V on the reveres current is dominated by the reverse current flowing in defect-induced pre-breakdown sites. By reaching the hard pre-breakdown voltage (here −13 V) the reverse current of the whole cell is dominated by the hard pre-breakdown reverse current and the defect-induced pre-breakdown current. Secondly, all pre-breakdown currents are very localized. Therefore, the reverse $I-V$ characteristic measured at the whole solar cell is not meaningful to decide whether there are harmful hot spots in the cell or not. It may happen that the reverse current flows through many weak pre-breakdown sites, then the current at each pre-breakdown site is low and may be not dangerous. However, if the whole reverse current is flowing only through a few pre-breakdown sites, these sites may heat up to temperatures which may damage the solar cell. Hence, an imaging method should be used to image the distribution of the reverse current (i.e. the number of pre-breakdown sites in a solar cell), to get a better criterion for the selection of off-specification solar cells. For this lock-in thermography is not necessary. It is sufficient if an IR image is taken immediately after applying a reverse bias.

Chapter 4

Summary

The aim of this work was to show the physical origins of non-idealities in $I-V$ characteristics of multicrystalline silicon solar cells. Non-idealities in forward $I-V$ characterisitics of silicon solar cells have many different origins. Linear shunts are one of the most harmful failures decreasing the efficiency of solar cells. Non-linear shunts and high series resistances have also an influence on the $I-V$ characteristics and may affect the efficiency of solar cells as well. The main topic of this work was the investigation of the breakdown behavior, i.e. the reverse $I-V$ characteristics, of solar cells. Breakdown in solar cells appear at unexpectedly low reverse voltages, and the formation of hot spots due to breakdown may damage solar cells. Since the breakdown behavior of solar cells was only rarely investigated in the past, the aim of this work was to reveal the origins of low breakdown voltages of solar cells. This summary gives a brief overview on the effects mentioned above starting with the forward $I-V$ characteristic.

Non-idealities in forward $I-V$ characteristics are caused by linear and non-linear shunts. A linear shunt appears if the n^+ emitter of the solar cell and the metallic back contact of the solar cell are in ohmic contact. Consequently, the light-generated current flows through the shunt channel and can not be used for power extraction from the solar cell. Therefore, linear shunts have to be avoided. There are a lot of different mechanisms leading to linear shunts. One has to distinguish between process-induced linear shunts, which are caused somewhere during the production process of the solar cells, and material-induced linear shunts, which are caused due to intrinsic defects or precipitates in the multicrystalline Si material used.

Process-induced linear shunts are the following ones: If the p-n junction at the edge of the solar cell is not separated correctly, a short circuit appears at the edge leading to an ohmic edge shunt. An emitter layer may be formed in or metal may be pressed through cracks in the Si wafer and therefore cause ohmic shunts. Another type of ohmic shunts are aluminum particles, which reach the emitter of the solar cell during the production process. The aluminum may form an ohmic contact to the p-doped base and a tunnel contact to the n^+ emitter during the firing step. Material-induced linear shunts are caused due to SiC precipitation in multicrystalline Si material. Typical SiC precipitation are SiC filaments found in grain boundaries of the mc-Si. These filaments may grow some millimetres in length towards the solidification direction of the Si ingot. Hence, they may penetrate the Si wafers used for solar cells production. Since it was found that SiC filaments are highly conductive, they may short circuit the emitter and the back contact and cause severe linear shunts. It was also observed that some SiC filaments cause weaker or even no linear shunts. Infrared microscopy investigations in this work revealed that this may be due to the morphology of the SiC filaments, if SiC filaments stuck in the wafer or branche out. Even if the filaments cross the wafer, in many cases they are also not producing shunts. The exact mechanism behind this phenomenon is not known yet. The growth of the SiC filaments is also not understood in detail until now. In this work it was demonstrated that SiC filaments show imperfect crystallographic structures, and it was assumed that they may grow due to solid state diffusion of carbon in Si. However, the calculation revealed that solid state diffusion is not sufficient to transport the necessary amount of carbon to the grain boundary, except if one assumes enhanced diffusion due to silicon self-interstitials, or one has to assume diffusion of the carbon in the grain boundaries of the mc-Si, which should be much faster than solid state diffusion. The validity of the last assumptions is not proven yet, maybe one has to assume the model of Möller [75] also for the SiC filaments in GB. This model assumes that the SiC filaments grow at the liquid–solid interface during the block-casting process. Also a combination of both models is feasible.

The impact of linear shunts on the $I-V$ characteristic of the solar cells are taken into account if the two diodes model is extended by the parallel resistance as shown in equation (1.5). The impact of non-linear shunts on the $I-V$ characteristic leads to increased current at voltage up to about 0.4 V. Two main types of non-linear shunts are known: Non-linear shunts due to recombination active extended defects crossing the space charge region, and non-linear shunts due to fired-through metallization on the front side of the cell, so called Schottky-type non-linear shunts. Scratches on the emitter of the cell and the edges of solar cells often show non-linear shunts. These non-linear shunts are due to enhanced recombination currents, which are caused by high defect densities at these sites. Scratches and edges show significantly higher ideality factors than n = 2 predicted by the established theory. By using a coupled defect level recombintion model these high ideality factors can be simulated and some conclusions on the nature of the defects are possible.

The real $I-V$ characteristics of reverse biased solar cells can not be explained by the established theory. The origins of breakdown occuring at much lower reverse voltages than expected from theory was shown in this work. For the investigation of reverse $I-V$ characteristics of solar cells new lock-in thermography methods were developed in the course of this work. The aim was to image the temperature coefficient of the reverse current and its slope. Therefore TC-DLIT and slope-DLIT were invented. TC-DLIT is used to image the distribution of the temperature coefficient in percent per Kelvin. Slope-DLIT is used to image the relative slope of the reverse current in solar cells (%/V). For both imaging methods it is only necessary to take standard DLIT images at various temperatures and voltages. Furthermore, an ILIT technique was developed to image the multiplication factor of avalanche breakdown. With this so-called MF-ILIT method it is possible to image directly the sites where avalanche multiplication in solar cells occurs.

Three ranges of pre-breakdown in solar cells can be distinguished: early pre-breakdown, defect-induced pre-breakdown, and hard pre-breakdown. Early pre-breakdown already occurs at voltages of about –4 V. The TC of early pre-breakdown is strongly negative but becomes less negative with increasing temperatures. Early pre-breakdown does not show avalanche multiplication. The reverse $I-V$ characteristics of early pre-breakdown sites are linear. The physical origins of the early pre-brakdown is still unclear.

Defect-induced pre-breakdown occurs at voltages from about –7 V. It shows negative TC at low temperatures and voltages, but their TC changes at higher temperatures and voltages towards almost zero or even positive values. Defect-induced pre-breakdown has almost exponential $I-V$ characteristics and does not show strong avalanche multiplication. The slope is moderate and keeps constant over nearly the whole temperature and voltage range investigated in this work. The origin of defect-induced pre-breakdown sites are recombination-active defects in the mc-Si material.

The breakdown mechanism is assumed to be a mixture of trap-assisted tunneling and trap-assisted avalanche.

Hard pre-breakdown occurs in acidically texturized solar cells at reverse voltages of about −13 V. The TC of hard pre-breakdown sites is strongly negative and their slope is significantly higher compared to that of early-, and defect-induced pre-breakdown sites. The TC becomes less negative with increasing temperatures. Hard pre-breakdown sites show an avalanche multiplication factor up to 4. The avalanche MF is, according to the avalanche mechanism, strongly temperature dependent and decreases with increasing temperature. The $I-V$ characteristics of hard pre-breakdown sites show a steep exponential slope at voltages at about −13 V. The physical origin of hard pre-breakdown is field enhancement at the bent p-n junction. The bending of the p-n junction is due to cone-shaped etch pits at defects, which are formed in the mc-Si material during the texturization process of solar cells. Hard pre-beakdown is observable only at etch pits which are formed at clean defects not showing recombination. However, etch pits may also occur at recombination-active defects, but there defect-induced pre-breakdown sites appear already at lower voltages.

Numerous physical origins of defect-induced non-idealities have been shown in this work. However, a lot of open questions still remain and new questions arise: for instance, what is the origin of the homogenous current which was observed in revers biased solar cells, what is the origin of the early pre-breakdown, and what is the true growth mechanism of SiC filaments in mc-Si material? Furthermore it is not clear yet which defects cause the cone-shaped holes leading to hard pre-breakdown sites. A lot of effort still has to be done to reveal all features influencing the $I-V$ characteristics of solar cells, affecting their efficiency and the yield of solar cells in solar cell production. It should be emphasized here that shunt currents as well as recombination currents and currents at pre-breakdown sites are local currents. From this it follows that some theoretical predictions do not hold for the determination of theoretical $I-V$ characteristics, because they are made for homogenous currents flowing in solar cells.

Appendix A

2D-Image Sets

A.1 J-images, TC-, Slope-DLIT, and MF-ILIT Images

Here the current density images (J-images), the temperature coefficient DLIT images, the slope-DLIT images, and the multiplication factor ILIT images of solar cell 1 are shown. As mentioned in section 3.2.1 each type of pre-breakdown is marked by a different circle. Exemplarily one early pre-breakdown site is marked by a dashed circles (three dashes), one defect-induced pre-break down site is marked by a solid circle, and one hard pre-breakdown site is marked by a dotted circle, respectively. The solar cell displayed in the images has an area of 156×156 mm^2.

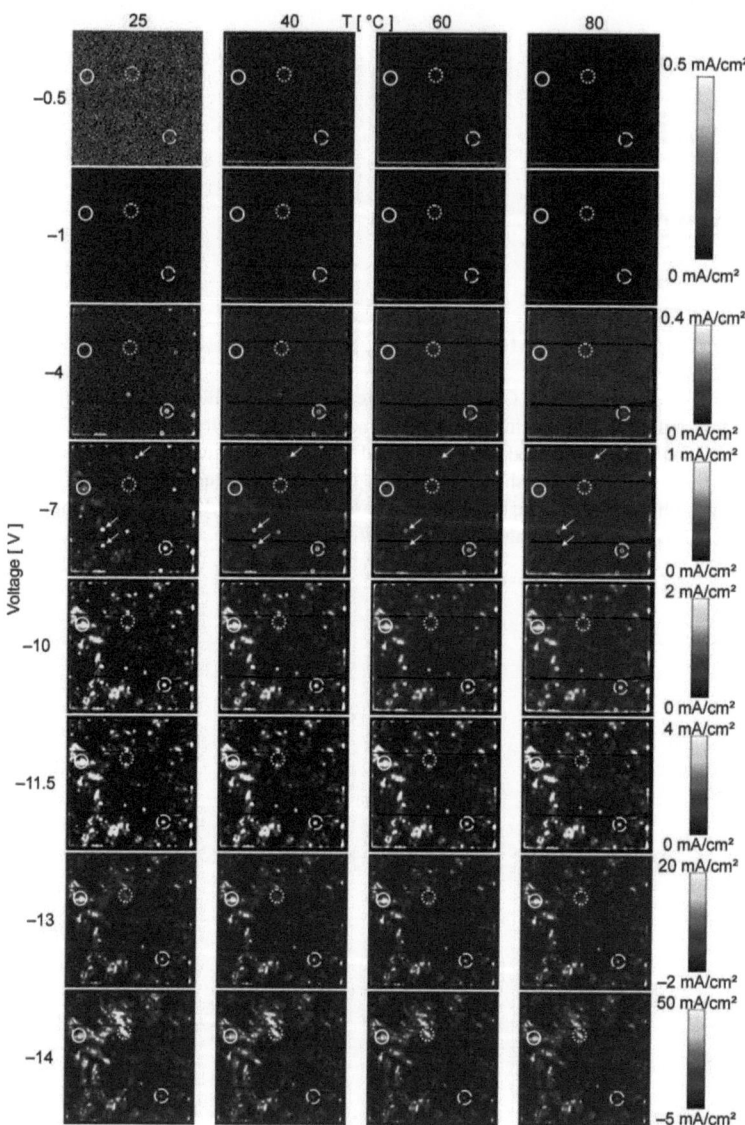

Figure A.1: 2-dimensional array of current density images of solar cell 1. Dashed circles: early pre-breakdown site, solid circles: defect-induced pre-breakdown site, dotted circles: hard pre-breakdown site. The arrows in image (25 °C versus –7 V) mark early-prebreakdown sites which break down somewhere between –4 V and –7 V reverse voltage.

A.1. J-IMAGES, TC-, SLOPE-DLIT, AND MF-ILIT IMAGES

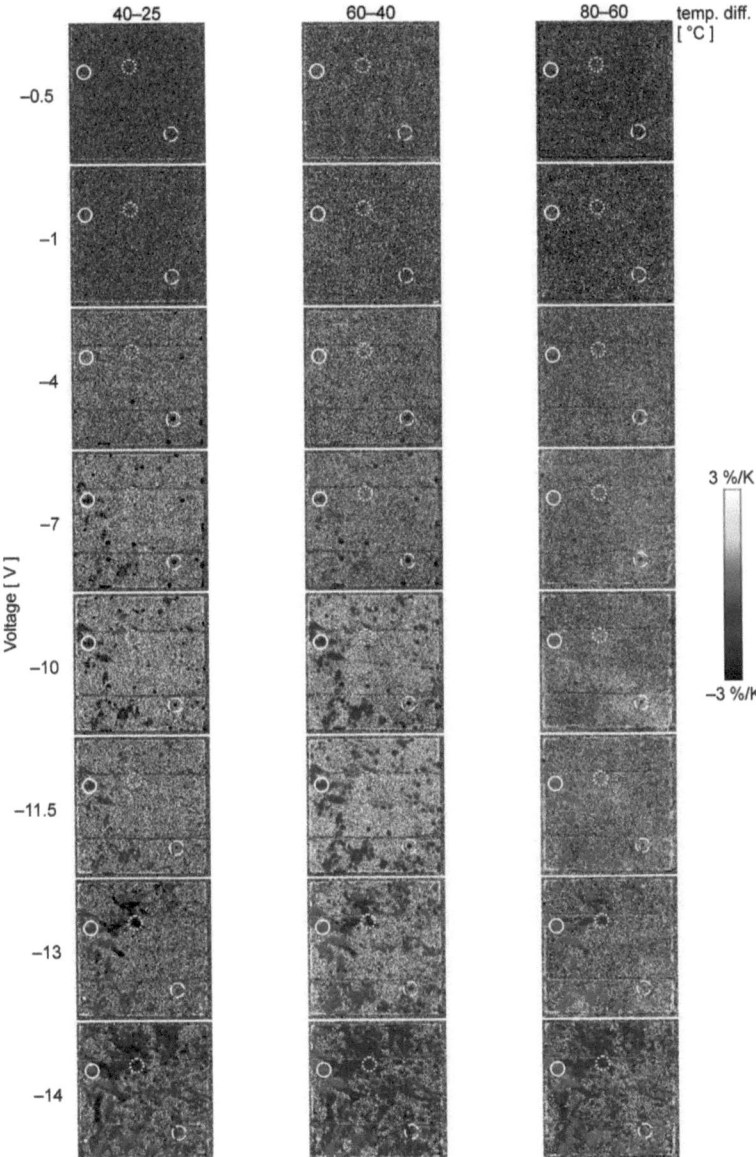

Figure A.2: 2-dimensional array of TC-DLIT images of solar cell 1. Dashed circles: early pre-breakdown site, solid circles: defect-induced pre-breakdown site, dotted circles: hard pre-breakdown site.

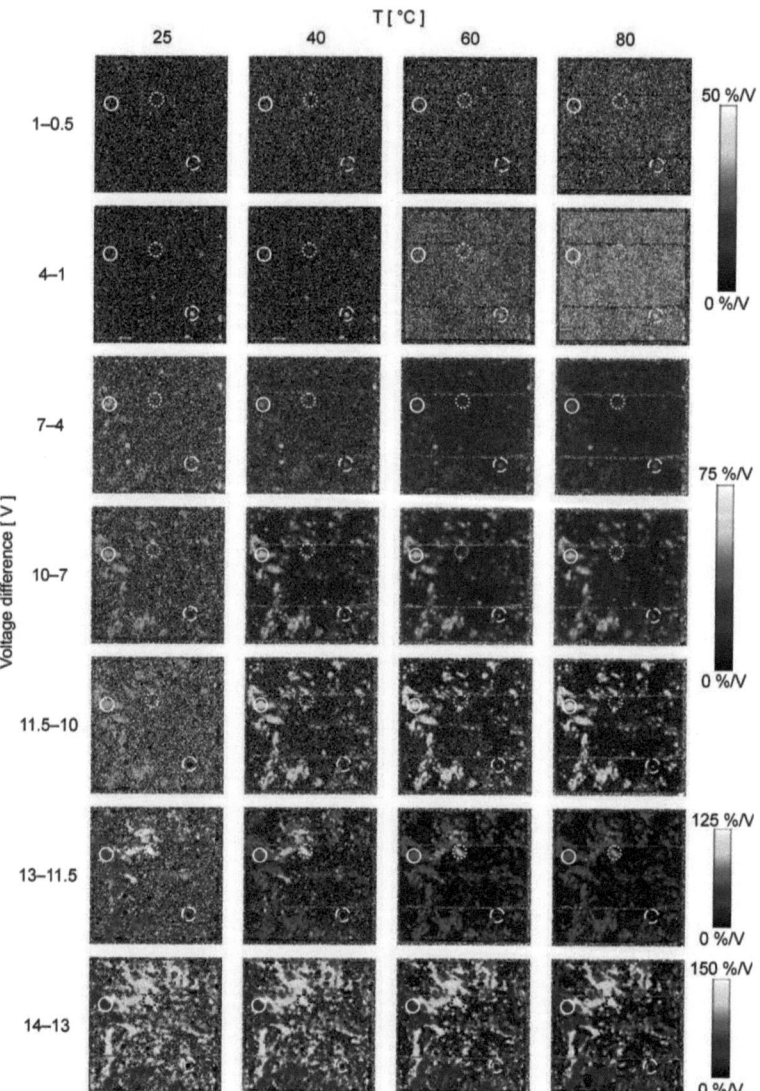

Figure A.3: 2-dimensional array of slope-DLIT images of solar cell 1. Dashed circles: early pre-breakdown site, solid circles: defect-induced pre-breakdown site, dotted circles: hard pre-breakdown site.

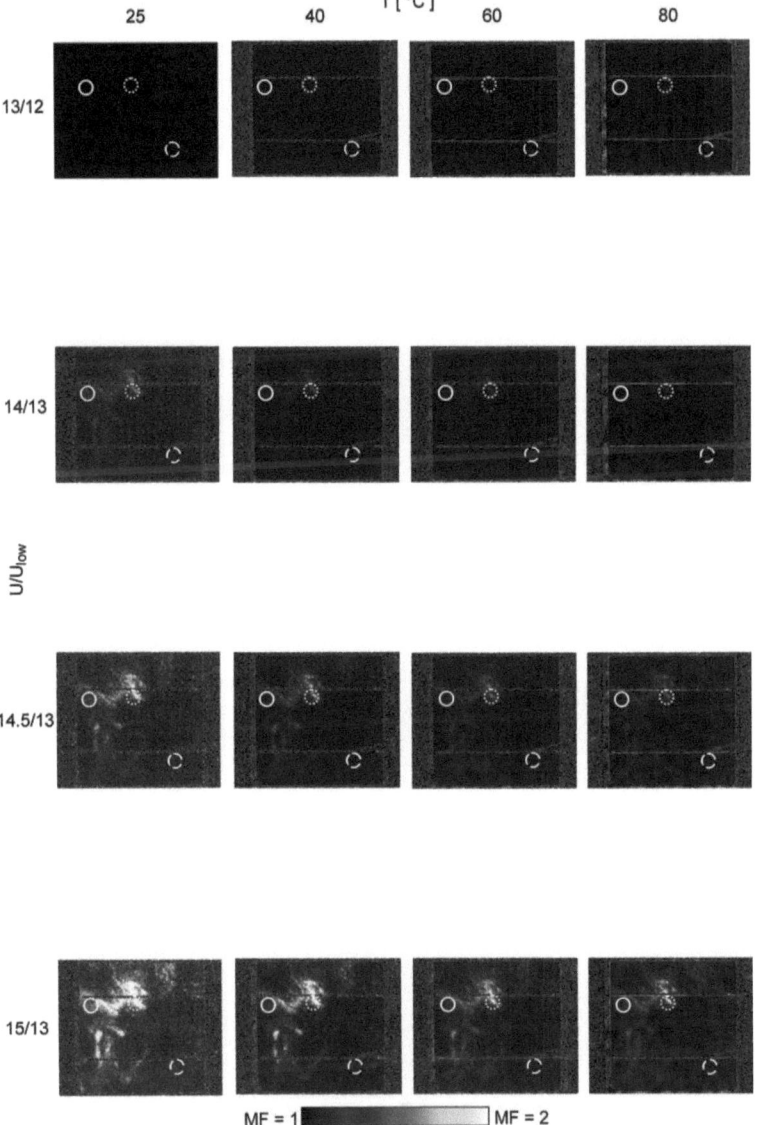

Figure A.4: 2-dimensional array of MF-ILIT images of solar cell 1. Dashed circles: early pre-breakdown site, solid circles: defect-induced pre-breakdown site, dotted circles: hard pre-breakdown site.

Bibliography

[1] P. Würfel, *Physics of Solar Cells*, Wiley-VCH Verlag GmbH & Co. KGaA, Weinheim (2005)

[2] M.A. Green, *Solar Cells - Operating Principles*, Technology and System Applications, The University of New South Wales (1998)

[3] A. Goetzberger, *Sonnenenergie: Photovoltaik*, B.G. Teubner, Stuttgart (1997)

[4] J. Dietl, D. Helmreich, and E. Sirtl, *"Solar" Silicon*, in: *Crystals: Growth, Properties, and Applications 5: Silicon*, Springer-Verlag, Berlin, Heidelberg, New York (1981), 43

[5] P.W. Bridgman, *Certain Physical Properties of Single Crystals of Tungsten, Bismuth, Tellurium, Cadmium, Zinc, and Tin*, Proceedings of the American Acadamy of Arts and Sciences **60** (1925), 305

[6] D.C. Stockbarger, *The Production of Large Single Crystals of Lithium Fluoride*, Review of Scientific Instruments **7** (1936), 133

[7] C.P. Khattak and F. Schmid, *Casting Large Silicon-Crystals in Clear Silica Crucibles*, American Ceramic Society Bulletin **57** (1978), 609

[8] T.F. Cisek, G.H. Schwuttke, and K.H. Yang, *Directionally Solidified Solar-Grade Silicon Using Carbon Crucibles*, Journal of Crystal Growth **46** (1979), 527

[9] D.H. Macdonald, A. Cuevas, M.J. Kerr, C. Samundsett, D. Ruby, and S. Winderbaum, A. Leo, *Texturizing Industrial Multicrystalline Silicon Solar Cells*, Solar Energy **76** (2004), 277

[10] W. Shockley, *The Theory of p-n Junctions in Semiconductors and p-n Junction Transistors*, Bell System Technical Journal **28** (1949), 435

[11] W. Shockley and W.T. Read, jr., *Statistics of the Recombinations of Holes and Electrons*, Physical Review **87** (1952), 835

[12] R.N. Hall, *Electron-Hole Recombination in Germanium*, Physical Review **87** (1952), 387

[13] C.T. Sah, R.N. Noyce, and W. Shockley, *Carrier Generation and Recombination in p-n Junction and p-n Junction Characteristics*, In Proceedings of the Institute of Radio Engineers **45** (1957), 1228

[14] K.R. McIntosh, *Lumps, Humps and Bumps: Three Detrimental Effects in the Current-Voltage Curve of Silicon Solar Cells*, PhD Thesis, University of New South Wales, Sydney (2001)

[15] S. Mahadevan, S.M. Hardas, and G. Suryan, *Electrical Breakdown in Semiconductors*, physica status solidi (a) **8** (1971), 335

[16] S.M. Sze and G. Gibbons, *Effect of Junction Curvature on Breakdown Voltage in Semiconductor*, Solid-State Electronics **9** (1966), 831

[17] W. Mönch, *On the Physics of Avalanche Breakdown in Semiconductors*, physica status solidi **36** (1969), 9

[18] C. Zener, *A Theory of the Electrical Breakdown of Solid Dielectrics*, Proceedings of the Royal Society A **145** (1934), 523

[19] S.M. Sze and K. K. Ng, *Physics of Semiconductor Devices*, John Wiley & Sons, Inc., Hoboken, New Jersey (2007)

[20] G.A.M. Hurkx, D.B.M. Klaassen, and M.P.G. Knuvers, *A New Recombination Model for Device Simulation*, IEEE Transactions on Electron Devices **39** (1992), 331

[21] A.G. Chynoweth and K.G. McKay, *Internal Field Emission in Silicon p-n Junctions*, Physical Review **106** (1957), 418

[22] K.B. McAfee, E.J. Ryder, W. Shockley, and M. Sparks, *Observations of Zener Current in Germanium p-n junctions*, Physical Review **83** (1951), 650

[23] H.J. Queisser, *Forward Characteristics and Efficiencies of Silicon Solar Cells*, Solid-State Electronics (1962) **5**, 1

[24] A. Kaminski, J.J. Marchand, H. El Omari, and A. Laugier, *Conduction Processes in Silicon Solar Cells*, In Proceedings of 25th IEEE Photovoltaics Specialists Conference, Washington D.C., USA (1996), 573

[25] A. Schenk and U. Krumbein, *Coupled Defect-Level Recombination: Theory and Application to Anomalous Diode Characteristics*, Journal of Applied Physics **78** (1995), 3185

[26] O. Breitenstein and J. Heydenreich, *Non-Ideal I–V Characteristics of Block-Cast Silicon Solar Cells*, Solid State Phenomena **37-38** (1994), 139

[27] R. Kühn, P. Fath, and E. Bucher, *Effects of pn-Junctions Boardering on Surfaces Investgated by Means of 2D-Modelling*, In Proceedings of 28th IEEE Photovoltaic Specialists Conference, Anchorage, USA (2000), 116

[28] H.S. Rauschenbach and E.E. Maiden, *Breakdown Phenomena in Reverse Biased Solar Cells*, In prodeedings of the 9th IEEE Photovoltaic Specialists Conference, Silver Springs, USA (1972), 217

[29] J.W. Bishop, *Computer-Simulation of the Effects of Electrical Mismatches in Photovoltaic Cell Interconnection Circuits*, Solar Cells **25** (1988), 73

[30] P. Spirito and V. Abergamo, *Reverse Bias Power Dissipation of Shadowed or Faulty Cells in Different Array Configurations*, In Proceedings of 4th European Photovoltaic Solar Energy Conference, Stresa, Italy (1982), 296

[31] R.A. Hartman, J.L. Prince, and J.W. Lathrop, *Second Quadrant Effect in Silicon Solar Cells*, In Proceedings of 14th IEEE Photovoltaic Specialists Conference, San Diego, USA (1980), 119

[32] M.C. Alonso-García and J.M. Ruíz, *Analysis and Modelling the Reverse Characteristic of Photovoltaic Cells*, Solar Energy Materials & Solar Cells **90** (2006), 1105

[33] V. Quaschning and R. Hanitsch, *Numerical Simulation of Current-Voltage Characteristics of Photovoltaic Systems with Shared Solar Cells*, Solar Energy **56** (1996), 513

[34] S.L. Miller, *Ionization Rates for Holes and Electrons in Silicon*, Physical Review **105** (1957), 1246

[35] J.M. Ruíz and M.C. Alonso-García, *Current Multiplication in Reverse Biased Silicon Solar Cells*, In Proceedings of 19th European Photovoltaic Solar Energy Conference, Paris, France (2004), 340

[36] J.W. Bishop, *Microplasma Breakdown and Hot-Spots in Silicon Solar Cells*, Solar Cells **26** (1989), 335

[37] A. Simo and S. Martinuzzi, *Hot Spots and Heavily Dislocated Regions in Multicrystalline Silicon Solar Cells*, In Proceeding of 21st IEEE Photovoltaic Specialists Conference, Kissimmee, USA (1990), 800

[38] O. Breitenstein and M. Langenkamp, Series in Advanced Microelectronics 10: *Lock-In Thermography, Basics and Use for Diagnostics of Electronic Components*, Springer-Verlag, Berlin-Heidelberg (2003)

[39] http://www.thermosensorik.de

[40] O. Breitenstein, J.P. Rakotoniaina, and M. Hejjo Al Rifai, *Quantitative Evaluation of Shunts in Solar Cells by Lock-In Thermography*, Progress in Photovoltaics: Research and Applications **11** (2003), 515

[41] O. Breitenstein, J.P. Rakotoniaina, M. Hejjo Al Rifai, and M. Werner, *Shunt Types in Crystalline Silicon Solar Cells*, Progress in Photovoltaics: Research and Applications **12** (2004), 529

[42] J. Isenberg and W. Warta, *Realistic Evaluation of Power Losses in Solar Cells by Using Thermographic Methods*, Journal of Applied Physics **95** (2004), 5200

[43] M. Kaes, S. Seren, T. Pernau, and G. Hahn, *Light-modulated Lock-In Thermography for Photosensitive pn-Structures and Solar Cells*, Progress in Photovoltaics: Research and Applications **12** (2004), 355

[44] O. Breitenstein, J.P. Rakotoniaina, A.S.H. van der Heide, and J. Carstensen, *Series Resistance Imaging in Solar Cells by Lock-In Thermography*, Progress in Photovoltaics: Research and Applications **13** (2005), 645

[45] K. Ramspeck, K. Bothe, D. Hinken, B. Fischer, J. Schmidt, and R. Brendel, *Recombination Current and Series Resistance Imaging of Solar Cells by Combined Luminescence and Lock-In Thermography*, Applied Physics Letters **90** (2007), 153502

[46] O. Breitenstein, J. Bauer, J.-M. Wagner, and A. Lotnyk, *Imaging Physical Parameters of Pre-Breakdown Sites by Lock-In Thermography Techniques*, Progress in Photovoltaics: Research and Applications **16** (2008), 679

[47] I. E. Konovalov, O. Breitenstein, and K. Iwig, *Local Current-Voltage Curves Measured Thermally (LIVT): A New Technique of Characterizing PV Cells*, Solar Energy Materials and Solar Cells **48** (1997), 53

[48] T. Fuyuki, H. Kondo, T. Yamazaki, Y. Takahashi, and Y. Uraoka, *Photographic surveying of minority carrier diffusion length in polycrystalline silicon solar cells by electroluminescence*, Applied Physics Letters **86** (2005), 262108

[49] T. Trupke, R. A. Bardos, M. C. Schubert, and W. Warta, *Photoluminescence Imaging of Silicon Wafers*, Applied Physics Letters **89** (2006), 044107

[50] O. Breitenstein, J. Bauer, T. Trupke, and R.A. Bardos, *On The Detection of Shunts in Silicon Solar Cells by Photo- and Electroluminescence Imaging*, Progress in Photovoltaics: Research and Applications **16** (2008), 325

[51] M. Kasemann, D. Grote, B. Walter, W. Kwapil, T. Trupke, Y. Augarten, R.A. Bardos, E. Pink, M.D. Abbott, and W. Warta, *Luminescence Imaging for the Detection of Shunts on Silicon Solar Cells*, Progress in Photovoltaics: Research and Applications **16** (2008), 297

[52] D. Hinken, K. Ramspeck, K. Bothe, B. Fischer, and R. Brendel, *Series Resistance Imaging of Solar Cells by Voltage Dependent Electroluminescence*, Applied Physics Letters **91** (2007), 182104

[53] H. Kampwerth, T. Trupke, J.W. Weber, and Y. Augarten, *Advanced Luminescence Based Effective Series Resistance Imaging of Silicon Solar Cells*, Applied Physics Letters **93** (2008), 202102

[54] F. Dreckschmidt, T. Kaden, H. Fiedler, and H.J. Möller, *Electroluminescence Investigation of the Decoration of Extended Defects in Multicrystalline Silicon*, In Proceedings of 22nd European Photovoltaic Solar Energy Conference, Milan, Italy (2007), 283

[55] R. Newman, *Visible Light from a Silicon p-n Junction*, Physical Review **100** (1955), 700

[56] T. Figielski and A. Toruń, *On the Origin of Light Emitted from Reverse Biased p-n junctions*, In Proceedings of 6th International Conference of Physics of Semiconductors, Exceter, United Kingdom (1962), 863

[57] S. Yamada and M. Kitao, *Recombination Radiation as Possible Mechanism of Light Emission from Reverse-Biased p-n Junctions under Breakdown Condition*, Japanese Journal of Applied Physics **32** (1993), 4555

[58] K. Bothe, P. Pohl, J. Schmidt, T. Weber, P. Altermatt, B. Fischer, and R. Brendel, *Electroluminescence Imaging as an in-Line Characterization Tool for Solar Cell Production*, In Proceedings of 21st European Photovoltaic Solar Energy Conference, Dresden, Germany (2006), 597

[59] J.I. Hanoka and R.O. Bell, *Electro-Beam-Induced Currents in Semiconductors*, Annual Reviews Material Science **11** (1981), 353

BIBLIOGRAPHY

[60] O. Breitenstein, J. Bauer, A. Lotnyk, and J.-M. Wagner, *Defect induced non-ideal dark I–V characteristics of solar cells*, Superlattices and Microstructures **45** (2009), 182

[61] O. Breitenstein, J. Bauer, M. Kittler, T. Aguirov, and W. Seifert, *EBIC and Luminescence Studies of Defects in Solar Cells*, Scanning **30** (2008), 331

[62] J. Heydenreich, H. Blumtritt, R. Gleichmann, and H. Johansen, *Combined Application of SEM(EBIC) and TEM for the Investigation of the Electrical Activity of Crystal Defects in Silicon*, in: Scanning Electron Microscopy I, SEM Inc., AMF O'Hare, Chicago (1981), 351

[63] C. Funke, Technische Universität Bergakademie Freiberg, private communication

[64] J.-P. Rakotoniaina, O. Breitenstein, M. Werner, M. Hejjo Al-Rifai, T. Buonassisi, M.D. Pickett, M. Ghosh, A. Müller, and N. Le Quang, *Distribution and Formation of Silicon Carbide and Silicon Nitride Precipitates in Block-Cast Multicrystalline Silicon*, In Proceedings of 20th European Photovoltaic Solar Energy Conference, Barcelona, Spain (2005), 773

[65] A-K. Søiland, E.J. Øvrelid, T.A. Engh, O. Lohne, and J.K. Tuset, Ø. Gjerstad, *SiC and Si_3N_4 Inclusions in Multicrystalline Silicon Ingots*, Materials Science in Semiconductor Processing **7** (2004), 39

[66] L. Libioulle, Y. Houbion, and J.-M. Gilles, *Very sharp platinum tips for scanning tunneling microscopy*, Review of Scientific Instruments **66** (1995), 97

[67] www.nanotechnik.com

[68] J. Bauer, O. Breitenstein, and J.-P. Rakotoniaina, *Electronic activity of SiC precipitates in multicrystalline solar silicon*, physica status solidi (a) **204** (2007), 2190

[69] J. Bauer, Diploma Thesis, Martin-Luther-University Halle-Wittenberg and Max Planck Institut of Microstructure Physics Halle, Halle (Saale) (2006)

[70] M. Hejjo Al Rifai, O. Breitenstein, J.P. Rakotoniaina, M. Werner, A. Kaminski, and N. Le Quang, *Investigation of Material-Induced-Shunts in Block-Cast Multicrystalline Silicon Solar Cells by SiC Precipitates Filaments*, In Proceedings of 19th European Photovoltaic Solar Energy Conference, Paris, France (2004), 632

[71] H. Morkoç, S. Strite, G.B. Gao, M.E. Lin, B. Sverdlov, and M. Burns, *Large-Band-Gap SiC, III-V Nitride, and II-VI ZnSe-Based Semiconductor Device Technologies*, Journal of Applied Physics **73** (1994), 1363

[72] A. Lotnyk, J. Bauer, O. Breitenstein, and H. Blumtritt, *A TEM Study of SiC Particles and Filaments Precipitated in Multicrystalline Si for Solar Cells*, Solar Energy Materials & Solar Cells **92** (2008), 1236

[73] S. Köstner, J. Bauer, J.-M. Wagner, and O. Breitenstein, *3D Imaging of Precipitates Inside Block-Cast Silicon*, Submitted to 24th European Photovoltaic Solar Energy Conference, Hamburg, Germany (2009)

[74] S. Möller and H.J. Möller, *An Investigation of the Formation of Silicon Carbide Precipitates in Metallurgical Silicon*, In Proceedings of 9th European Photovoltaic Solar Energy Conference, Freiburg, Germany (1989), 439

[75] H.J. Möller, *Formation of Micro-Precipitates at the Melt-Solid Interface in Semiconductors*, In Proceedings of Material Research Society Symposium **205** (1991), 423

[76] F. Schmid, C.P. Khattak, T.G. Digges, Jr., and Larry Kaufman, *Origin of SiC Impurities in Silicon Crystals Grown from the Melt in Vacuum*, Journal of the Electrochemical Society: Electrochemical Science and Technology **126** (1979), 935

[77] W. Dietze, W. Keller, and A. Mühlbauer, *Float-Zone Grown Silicon*, in: *Crystals: Growth, Properties, and Applications 5: Silicon*, Springer-Verlag, Berlin, Heidelberg, New York (1981), 1

[78] T. Nozaki, Y. Yatsurugi, and N. Akiyama, *Concentration and Behavior of Carbon in Semiconductor Silicon*, Journal of the Electrochemical Society: Solid State Science **117** (1970), 1566

[79] A-K. Søiland, E.J. Øvrelid, O. Lohne, J.K. Tuset, T.A. Engh, and Ø. Gjerstad, *Carbon and Nitrogen Contents and Inclusion Formation During Crystallization of Multi-Crystalline Silicon*, In Proceedings of 19th European Photovoltaic Solar Energy Conference, Paris, France (2004), 911

[80] T.B. Massalski (Editor-in-Chief), *Binary Alloy Phase Diagrams*, ASM International (1990), 882

[81] J.P. Kalejs, L.A. Ladd, and U. Gösele, *Self-interstitial Enhanced Carbon Diffusion in Silicon*, Applied Physics Letters **45** (1984), 268

[82] H.J. Möller, C. Funke, J. Bauer, S. Köstner, H. Straube, and O. Breitenstein, *Growth of Silicon Carbide Filaments in Multicrystalline Silicon for Solar Cells*, Submitted to the Conference Gettering and Defect Engineering in Semiconductor Technology, Döllnsee-Schorfheide, Germany (2009), and private communication

[83] R. Newman and J. Wakefield, *The Diffusivity of Carbon in Silicon*, Journal of Physics and Chemisry of Solids **19** (1961), 230

[84] F. Rollert, N.A. Stolwijk, and H. Mehrer, *Diffusion of Carbon-14 in Silicon*, Materials Science Forum **38-41** (1989), 753

[85] R.F. Scholz, PhD Thesis, Martin-Luther-University Halle-Wittenberg, Halle (Saale) (1999)

[86] O. Breitenstein, P. Altermatt, K. Ramspeck, and A. Schenk, *The Origin of Ideality Factros n > 2 of Shunts and Surface in the Dark I–V Curves of Si Solar Cells*, In Proceedings of 21th European Photovoltaic Solar Energy Conference, Dresden, Germany (2006), 625

[87] O. Breitenstein, J. Bauer, P.P. Altermatt, and K. Ramspeck, *Influence of Defects on Solar Cell Characteristics*, Submitted to the Conference Gettering and Defect Engineering in Semiconductor Technology, Döllnsee-Schorfheide, Germany (2009)

[88] Sentaurus, TCAD, Synopsys Inc., Mountain View, California, USA (2005)

[89] W. Hermann, M. Adrian, and W. Wiesner, *Operational Behaviour of Commercial Solar Cells Under Reverse Biased Conditions*, In Proceedings of 2nd World Conference and Exhibition on Photovoltaic Solar Energy Conversion, Vienna, Austria (1998), 2357

[90] M.C. Alonso-García, W. Herrmann, W. Böhmer, and B. Proisy, *Thermal and Electrical Effects Caused by Outdoor Hot-spot Testing in Associations of Photovoltaic Cells*, Progress in Photovoltaics: Research and Applications **11** (2003), 293

[91] M.C. Alonso-García, J.M. Ruiz, and F. Chenloa, *Experimental Study of Mismatch and Shading Effects in the I–V Characteristic of a Photovoltaic Module*, Solar Energy Materials & Solar Cells **90** (2006), 329

[92] J. Munõz, E. Lorenzo, F. Martínez-Moreno, L. Marroyo, and M. García, *An Investigation into Hot-Spots in Two Large Grid-Connected PV Plants*, Progress in Photovoltaics: Research and Applications **16** (2008), 693

[93] D. Lausch, K. Petter, H. von Wenckstern, and M. Grundmann, *Correlation of Pre-Breakdown Sites and Bulk Defects in Multicrystalline Silicon Solar Cells*, Physica Status Solidi - Rapid Research Letters **3** (2009), 70

[94] W.K. Loke, S.F. Yoon, S. Wicaksono, K.H. Tan, and K.L. Lew, *Defect-Induced Trap-Assisted Tunneling Current in GaInNAs Grown on GaAs Substrate*, Journal of Applied Physics **102** (2007), 054501

[95] J. Bauer, J.-M. Wagner, A. Lotnyk, H. Blumtritt, B. Lim, J. Schmidt, and O. Breitenstein, *Hot Spots in Multicrystalline Silicon Solar Cells: Avalanche Breakdown due to Etch Pits*, Physica Status Solidi - Rapid Research Letters **3** (2009), 40

[96] D.V. Speeney and G.P. Carey, *Experimental Study of the Effect of Junction Curvature on Breakdown Voltage in Si*, Solid State Electronics **10** (1967), 177

Abbreviations

2D	2-dimensional
AB	avalanche breakdown
ARC	antireflection coating
BSF	back surface field
CCD	charge coupled device
DLIT	dark lock-in thermography
EBIC	electron beam-induced current
EFI	extended focus imaging
EL	electroluminescence
FF	fill factor
FIB	focused ion beam
GB	grain boundary
HF	hydrofluoric acid
HNO_3	nitric acid
I–V	current-voltage
IFE	internal field effect
ILIT	illuminated lock-in thermography
IRM	infrared microscopy
KOH	potassium hydroxide
LIT	lock-in thermography
mc	multicrystalline
MC	multiplication coefficient
MF	multiplication factor
PECVD	plasma-enhanced chemical vapor deposition
PL	photoluminescence
$POCl_3$	phosphorus oxychloride
SEM	scanning electron microscopy
SRH	Schockley-Read-Hall
TAA	trap-assisted avalanche
TAT	trap-assisted tunneling
TC	temperature coefficient
TEM	transmission electron microscopy

Acknowledgment

Zum Gelingen dieser Doktorarbeit habe ich viel Unterstützung und Hilfe erfahren, sei es fachlich, organisatorisch, emotional oder finanziell. An dieser Stelle möchte ich mich deswegen bei allen sehr herzlich bedanken, die, auf welche Weise auch immer, zum Gelingen meiner Doktorarbeit beigetragen haben. Da in einer Danksagung das Wort „danke" immer sehr häufig vorkommt (weshalb sich Danksagungen oft eigenartig lesen), möchte ich an dieser Stelle betonen, dass ich allen Personen, die in diesem Text namentlich genannt werden, aber auch vielen Personen, die in diesem Text nicht namentlich vorkommen, „Danke!" sage.

Herr Prof. Gösele gab mir dir Möglichkeit und das Vertrauen, an einem Institut mit hervorragender Ausstattung und allen Freiheiten, die man zum Forschen braucht, zu arbeiten. Außerdem hält er noch immer seine Hand schützend über die kleine, und am Max-Planck-Institut für Mikrostrukturphysik doch etwas fremde, Fachgruppe für Solarzellencharakterisierung.

Otwin „Otto" Breitenstein hat mir den Traum erfüllt, an Solarzellen zu forschen. Er hat mich mit seiner großartigen fachlichen Kompetenz und seinem menschlichen Umgang hervorragend in allen Belangen unterstützt, die für das Gelingen einer Doktorarbeit notwendig sind. Seine Begeisterung für Solarzellen, seine organisatorischen Fähigkeiten, Geld zu beschaffen, sein Vertrauen in meine Arbeit, die geduldige Beantwortung jeder Frage und die Möglichkeit, unsere Forschungsergebnisse auf Konferenzen zu präsentieren, haben mir die Arbeit an meiner Dissertation zu einer sehr wertvollen Erfahrung werden lassen.

Meine Arbeitsgruppe war für das Gelingen dieser Arbeit besonders wichtig. Auch wenn es seltsam anmuten mag, widme ich Jan-Martin Wagner für seine Hilfe bei theoretischen fachlichen, orthographischen und grammatikalischen Fragen eine extra Leerstelle in diesem Text. Hilmar Straube und mein ehemaliger Kollege Jean-Patrice Rakotoniaina diskutierten mit mir wichtige physikalische Fragen und vor allem Probleme mit der für uns so wichtigen Thermokamera. Stefan Köstner nahm mir viel Arbeit bei einem Thema ab, das ich selbst nicht mehr hätte so im Detail bearbeiten können.

Andriy Lotnyk und Nikolai Zakharov entdeckten viele Dinge am Transmissionselektronenmikroskop, die sehr wichtig für meine Arbeit waren. Im gleichen Zusammenhang muß Horst Blumtritt erwähnt werden, der für die Probenpräparation mit Hilfe des Focused Ion Beam sorgte. Herr Erfurth und Herr Aßmann sorgten dafür, dass die für mich wichtigen Elektronenmikroskope immer schöne Bilder lieferten. Von Pietro Altermatt und Klaus Ramspeck (ISFH Hameln) stammen die Simulationen des CDL-Modells, deren Ergebnisse ich dankenswerterweise hier zeigen durfte. Frau Hopfe und ihre Kolleginnen haben mir viele sehr nützliche Tipps gegeben, wie man Silicium mit Säuren malträtiert und sich in einem Chemielabor verhält, ohne lebensgefährlich verletzt zu werden!

Für den extrem reibungslosen Ablauf von Dingen hinter den Kulissen (Bezahlung, Posterdrucke, Bibliothek, Computer), von denen ich keine Ahnung habe, die aber sehr wichtig sind, müssen hier alle Mitarbeiterinnen und Mitarbeiter der Verwaltung und der Infrastruktur des Max-Planck-Instituts für Mikrostrukturphysik Halle besonders erwähnt werden. Ihr macht einen super Job! Natürlich dürfen auch die Elektronikwerkstatt und die Mechanische Werkstattt nicht unerwähnt bleiben, ohne deren kompetente Arbeit viele Ideen für Messungen und vor allem die zugehörigen Messgeräte undenkbar wären.

ACKNOWLEDGMENT

Für schöne Ablenkung von der Arbeit und für glücklicherweise nicht-fachliche Diskussionen beim Tages-AbschlußBier (TAB) danke ich allen Doktorandenkollegen, die daran teilgenommen haben. Außerdem muß die gute Arbeitsatmosphäre am Institut erwähnt werden.

Die Unterstützung meiner Familie, die meine Gesundheit und mein Wohlbefinden stets als wichtiger angesehen hat als das Gelingen eines meiner „verrückten" Experimente, empfand ich als besonders schön. Auch sollen meine Freunde nicht unerwähnt bleiben, die zum Glück ganz andere Sachen machen als ich und mich über den Tellerrand des Physikers hinaussehen lassen und mich dabei inspirieren. Einmal möchte ich doch ganz explizit das Wort „danke" benutzen, dies gilt meiner Freundin Doro, die mich sehr glücklich macht und mich oft daran erinnert, dass es neben p-n-Übergängen noch viel wichtigere Sachen im Leben gibt.

i want morebooks!

Buy your books fast and straightforward online - at one of world's fastest growing online book stores! Environmentally sound due to Print-on-Demand technologies.

Buy your books online at

www.get-morebooks.com

Kaufen Sie Ihre Bücher schnell und unkompliziert online – auf einer der am schnellsten wachsenden Buchhandelsplattformen weltweit! Dank Print-On-Demand umwelt- und ressourcenschonend produziert.

Bücher schneller online kaufen

www.morebooks.de

VDM Verlagsservicegesellschaft mbH
Heinrich-Böcking-Str. 6-8   Telefon: +49 681 3720 174   info@vdm-vsg.de
D - 66121 Saarbrücken   Telefax: +49 681 3720 1749 www.vdm-vsg.de

Printed by Books on Demand GmbH, Norderstedt / Germany